D1037194

ILLUSTRATED BASIC
CARPENTRY

ALSO BY *Graham Blackburn*

ILLUSTRATED HOUSEBUILDING

THE ILLUSTRATED ENCYCLOPEDIA OF WOODWORKING HANDTOOLS, INSTRUMENTS, & DEVICES

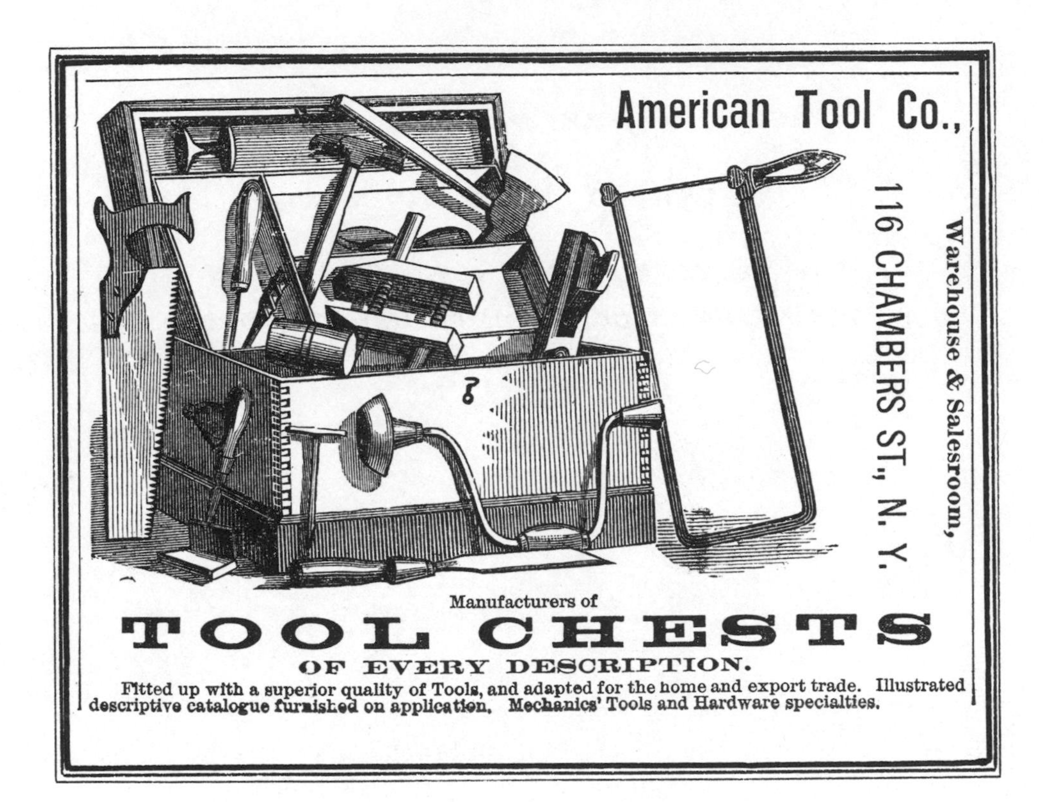

ADVERTISEMENTS FROM "THE IRON AGE" 1879

Illustrated Basic Carpentry

WRITTEN AND ILLUSTRATED BY

Graham Blackburn

BOBBS - MERRILL
INDIANAPOLIS · NEW YORK

copyright © 1976 by G.J. Blackburn

All rights reserved including the right of reproduction
in whole or in part in any form

Published by The Bobbs-Merrill Company, Inc.
Indianapolis · New York

ISBN 0-672-52064-8
LIBRARY OF CONGRESS CATALOG CARD NUMBER 74-17961

Designed by Graham Blackburn
Manufactured in the United States of America

First Printing

for
Oliver Duggal

"When a man teaches his son no trade, it is as if he taught him highway robbery."

TALMUD

Acknowledgments

I am grateful to the following people who made my life easier while I was writing this book: Heide Duggal for her love and care, Uta Rose, Barry Cahn for making the Cottage possible, Eugene Reynolds and Peter Walther for their help with it, Emma Ernst, alias Nina Newshop, for her breakfasts, and James-James Morrison-Brown for his companionship.

Although '**Carpentry**' is properly: "the art of shaping and assembling structural woodwork (as in constructing buildings)"* and the term '**Woodworking**' means: "the act, process, or occupation of working with wood,"* including carpentry, joinery, furniture-making, etc., the word '**Carpentry**' has been used in the title of this book, (although '**Woodworking**' would have been more exact,) since it is thought that common usage – the ultimate authority in the meaning of words – considers more and more that '**Carpentry**' now means what '**Woodworking**' used to, the more so with the gradual decline of, and loss of distinction between, the various occupations of carpenter, joiner, and cabinet-maker.

* WEBSTER'S THIRD NEW INTERNATIONAL DICTIONARY, 1966

Contents

PREFACE 9

CHAPTER ONE — Measuring Marking & Holding 13

MEASURING TOOLS 13, MARKING TOOLS 17, USING MEASURING & MARKING TOOLS 20, HOLDING TOOLS 27

CHAPTER TWO — Sawing 31

SAW TYPES 31, CROSSCUT SAWING 36, RIP SAWING 39, USING THE BACKSAW 41

CHAPTER THREE — Planing 43

PLANE TYPES 44, PARTS OF A PLANE 47, ADJUSTING A PLANE 48, USING A BENCH PLANE 53, USING THE BLOCK PLANE 57

CHAPTER FOUR *Cutting* 59

CUTTING TOOLS 59, HOW CUTTING WORKS 62, CHISELS 64, HOW TO USE CHISELS 67

CHAPTER FIVE *Boring* 73

BORING TOOLS 73, HOW TO USE DRILLS 77, HOW TO USE AUGER BITS & BIT BRACES 81

CHAPTER SIX *Striking* 85

STRIKING TOOLS 85, HAMMERING 88, USING MALLETS 94, TIGHTENING HANDLES 95

CHAPTER SEVEN *Tool Care & Sharpening* 97

GENERAL CARE 98, SHARPENING 100

BIBLIOGRAPHY 111

INDEX 115

Preface

Although the working of wood by hand has remained basically the same for several centuries, the use of power tools, introduced towards the end of the last century, has continued with ever increasing acceleration to the point now where much that is basic has either fallen into disuse or has been forgotten.

Now while few craftsmen would be willing to use only handtools, there are many who begin solely with power tools, expecting their exclusive use to be sufficient. This is not, however, the case. The ability to use handtools is fundamental to all woodworking practice.

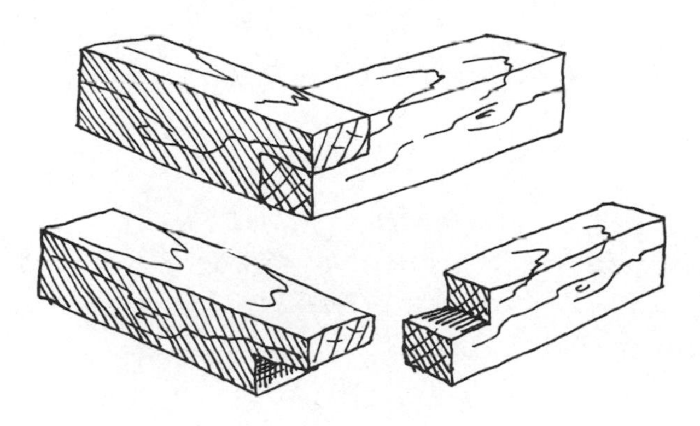

A HALVED-TOGETHER JOINT

Wood is fast becoming a precious commodity. Its use in the building industry, for example, is being constantly diminished, both by its ever-increasing price and the introduction of new materials such as plastic and aluminum. Its use in the future will have to be justified, therefore, not by its abundance and cheapness, but by its merits as a beautiful, if expensive and relatively scarce, material.

This has, indeed, already happened to a large extent. Where, previously, hardwoods such as walnut and mahogany were commonly used in joinery, as window and door frames, for example, now only pine is used; for walnut and mahogany have become so expensive that they may be used only by fine furniture makers and cabinet makers.

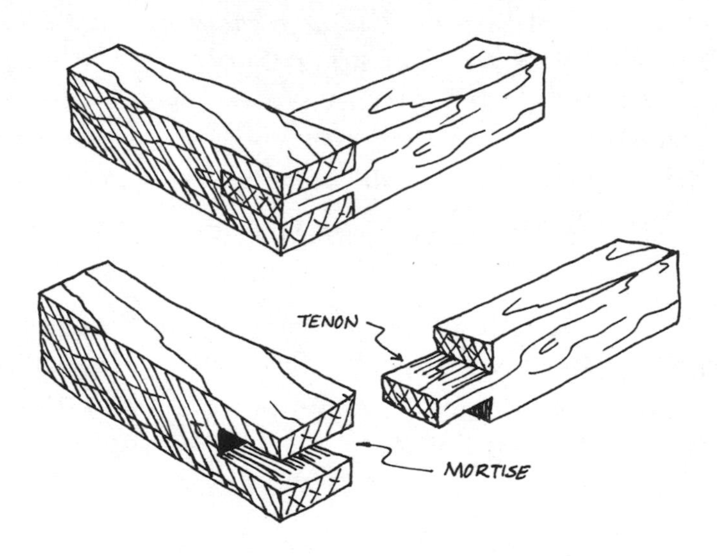

AN OPEN MORTISE-AND-TENON JOINT

To further point the process, where door frames were once commonly made from 5/4 stock (i.e. wood that is 1¼" thick) they are now, in all but the very best quality work, made from wood that is only ¾" thick.

In the very near future, faced with high prices and scarcity of materials, the wood butchery that now passes for a lot of woodwork, from carpentry to furniture making, will be unacceptable to most people.

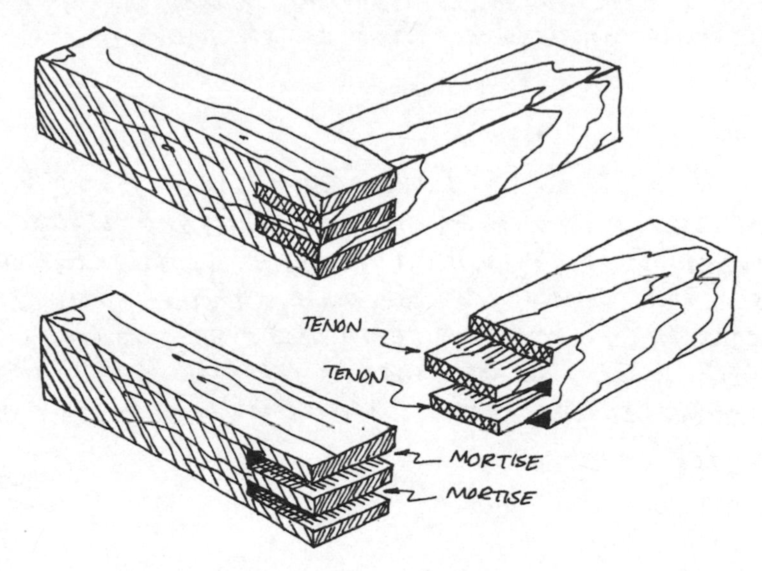

AN OPEN, DOUBLE MORTISE-AND-TENON JOINT

The respect necessary to use and work wood well and beautifully is developed more readily by the close contact obtained through the use of handtools than by the more impersonal sense of omnipotence which the use of power tools imparts. To sit on a piece of wood and work a moulding with a hand plane is to come to know that piece of wood almost as if it were a person. To thrust it through a noisy machine, in fear for your fingers, is to be totally unaware of its knots and its grain, its feeling and its texture.

This is not to decry, far less to prohibit, the use of power tools, but merely to point out that by learning to use handtools first, an understanding and a feeling for the material will be developed that is difficult to obtain across the scream of an electric table-saw. When, however, this feeling has been developed, then the table-saw will be able to be used to full advantage.

While true expertise comes only with practice, the proper use of handtools is not hard to learn. This book seeks to demonstrate how simple are the basic ingredients: sawing, planing, cutting, boring, and striking. These are the five fundamental operations. A separate book could be written about each of them (to which end I have included a bibliography covering more advanced aspects) but, in essence, that is all there is.

I have begun the book with some remarks on measuring and marking, for accuracy is essential to success. I have ended the book with a chapter on the care and sharpening of tools. This comes last because, although you cannot use any tool successfully unless it be in good order, it makes more sense to learn first what the tool may be expected to do, for then you will have a more sympathetic understanding of how and why it should be cared for.

A word about tools: buy only the best. It is better to have only a few tools of excellent quality than many of poor quality. There is no sense, while you are an inexperienced and second-class craftsman, in putting yourself at the double disadvantage of using second-class tools. They will only cost you more time and money in the long run. Treat your tools, your wood, and yourself with respect. Wood is a beautiful material, and the process of working it is as important as the result. Like all crafts, woodworking is as much an art as a trade. Most of all, be patient, for it is experience, not books, which teaches.

Graham Blackburn

WOODSTOCK, NEW YORK 1974

CHAPTER ONE

Measuring Marking & Holding —

"IN GOOD CARPENTRY EVERYTHING DEPENDS ON ACCURACY OF MEASUREMENT..." *from Elementary Carpentry, 1870*

There is very little in woodworking that is very complicated. It is mostly a question of performing basically simple operations with care. No matter how carefully you cut or plane, or saw or drill, if you have not measured and marked your work carefully at the start nothing will fit or be true and strong. This chapter tells you how to do this, and also explains the various holding tools — so that you can keep your work secure while doing everything from measuring to finishing.

MEASURING TOOLS The commonest tool used for measuring in woodwork is the **tape measure**. Formerly, **tape measures** were made with cloth tapes and were used chiefly for measuring long distances like 25' to 100'.

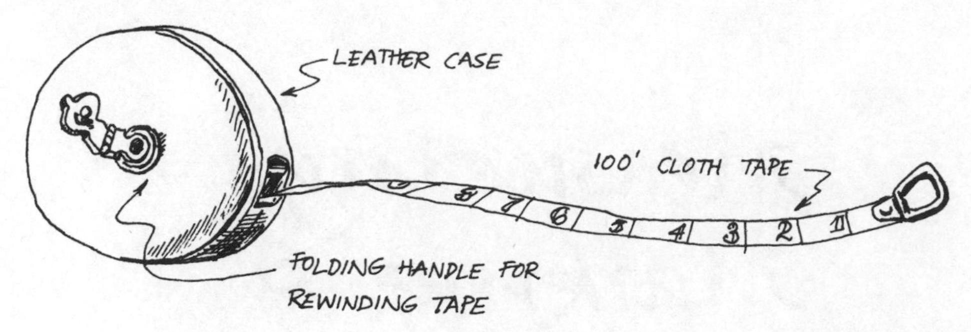

CLOTH TAPE MEASURE

Since cloth stretches, these tapes weren't completely accurate, and so now tapes are made of metal which doesn't stretch at all. **Steel tape measures** are used for short and long measurements. They are made in lengths from 6' to 100', but the commonest, and most useful, are the 12' and 16' lengths.

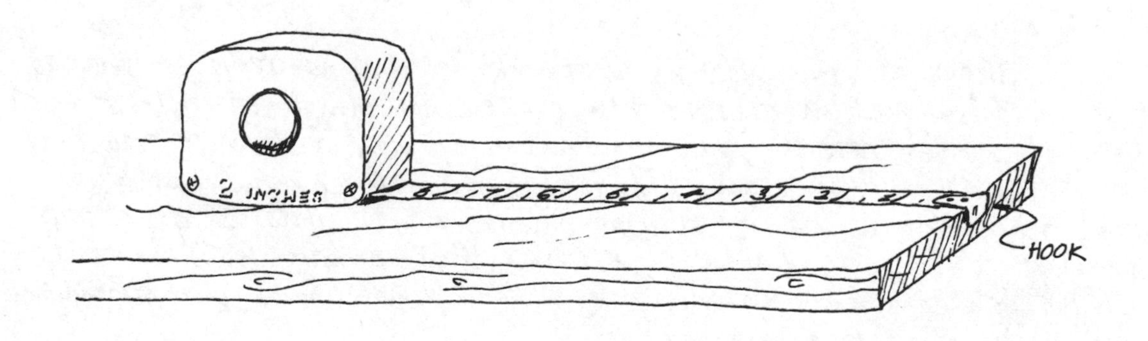

The better ones have hooked ends to hold on the edge of wood, and are spring loaded so that they retract automatically and do not need to be wound up as do **cloth tape measures.** The length of the bottom of the case is usually marked so that inside measurements can be read by reading the last figure visible on the tape and adding on the marked length of the case.

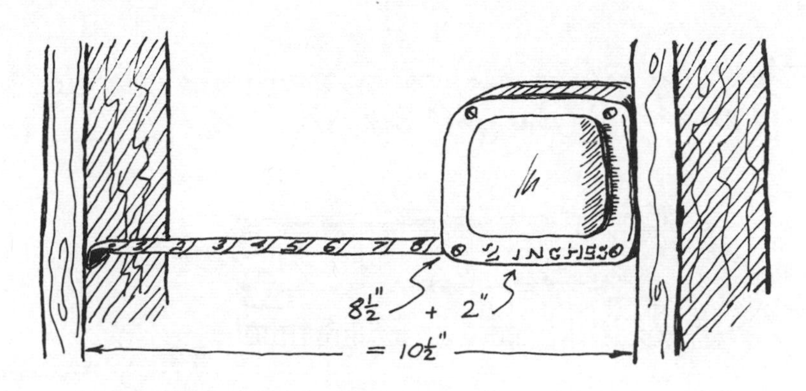

AN INSIDE MEASUREMENT

The next commonest **rule**, used especially by carpenters, is the **zig-zag folding-rule**. These usually unfold to 6' or 8', and sometimes even longer, but the 6' **rule** is the best to use, for they break easily when fully unfolded, and it takes a lot longer to fold up a **zig-zag folding-rule** than it does to rewind a metal **tape measure**, so the shorter the better. However, since both the **tape measure** and the **folding-rule** measure equally well, it is mostly a matter of personal preference which is used.

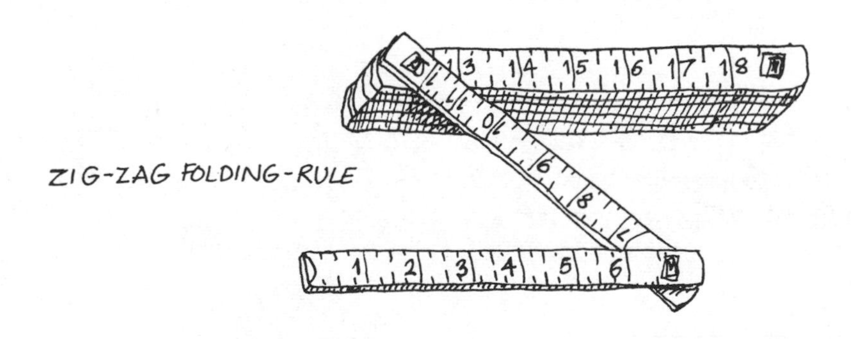

ZIG-ZAG FOLDING-RULE

For inside measurements, the better **zig-zag rules** have a sliding extension, the reading of which is added to the length of the opened **rule**.

The third general-purpose woodworking **rule** is the now somewhat old-fashioned **four-fold two-foot rule**.

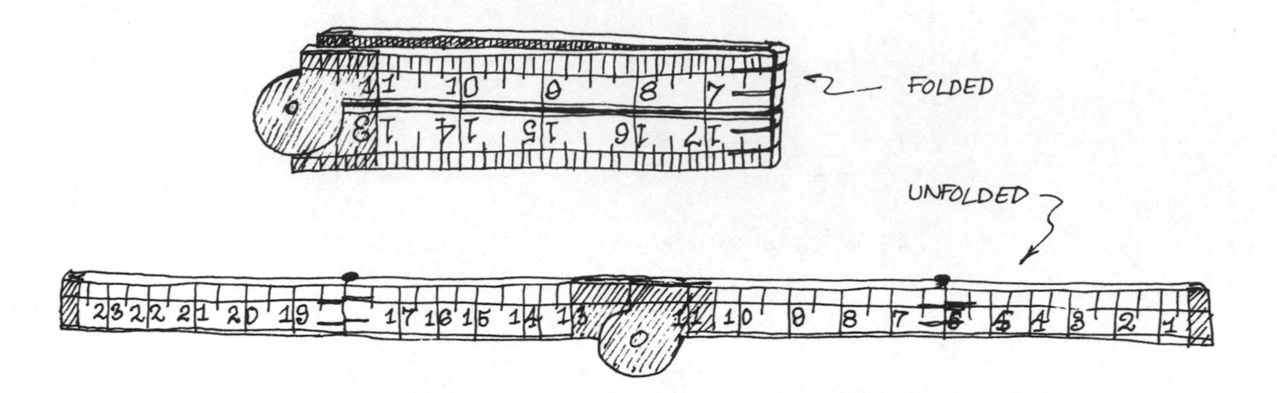

Although not as convenient as steel **tape measures**, since they take up more space and measure only 2' (although longer ones are occasionally made), many older woodworkers can use them with great dexterity. In fact, this type of **rule** used to be known as simply the **carpenter's rule**. Normally made of boxwood, it is graduated, like most other rules, in inches, ⅛ths, and 1/16ths of inches.

Lastly, there is the **bench rule**, used less by carpenters than by joiners or cabinetmakers. As its name implies, this is the **rule** used by woodworkers who work at a bench.

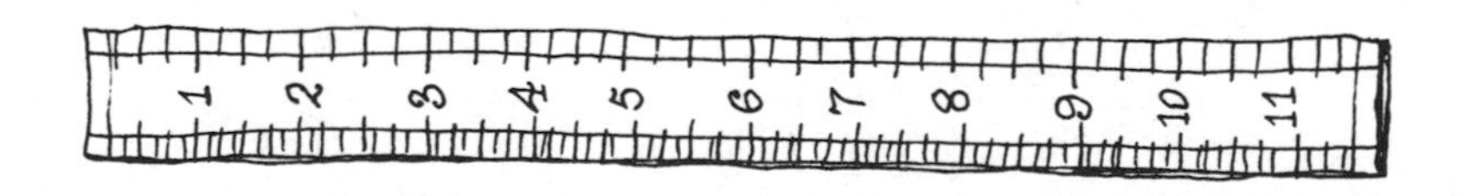

Usually a straight, and often metal-edged, wooden **rule**, it is used where measurements must be made with greater accuracy than is demanded by rough carpentry, and consequently is graduated in more detail, often to 1/100ths of an inch.

MARKING TOOLS

In order to be able to use the measuring tools you must be able to note what you have measured. This is usually done directly on the wood with various **pencils**, **knives**, **awls** or **scribers**.

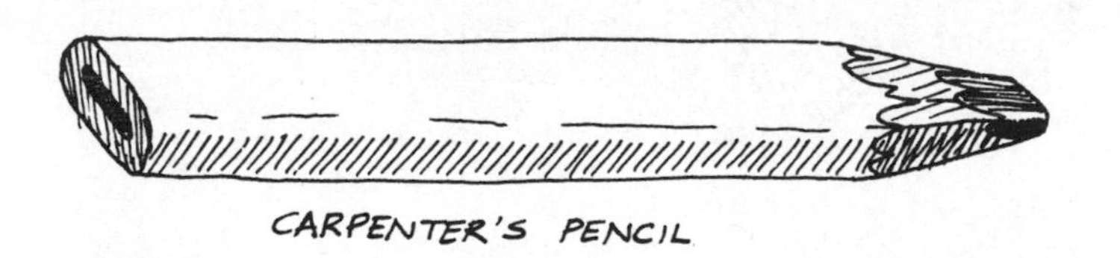

CARPENTER'S PENCIL

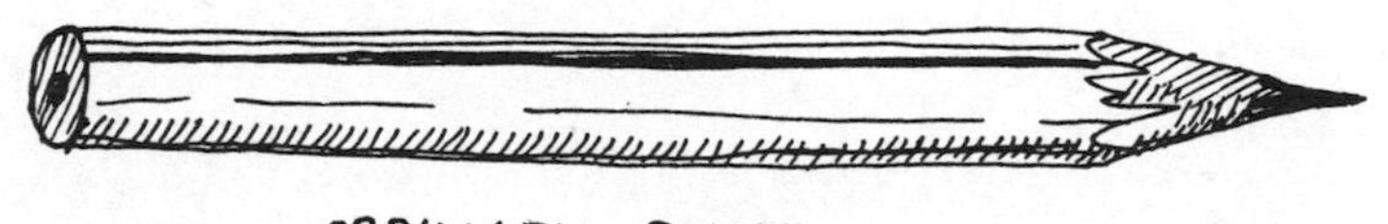

ORDINARY PENCIL

Pencils are the commonest markers. The **carpenter's pencil** with its broad, soft lead is used for rough general marking; **ordinary pencils**, with leads hard enough to hold a point, but not so hard as to scratch deeply, being used for more accurate lines.

Where lines of great accuracy or permanence are required, a **knife** or **scriber** is used. The kind of **knife** most used by woodworkers at their benches is the **sloyd knife**.

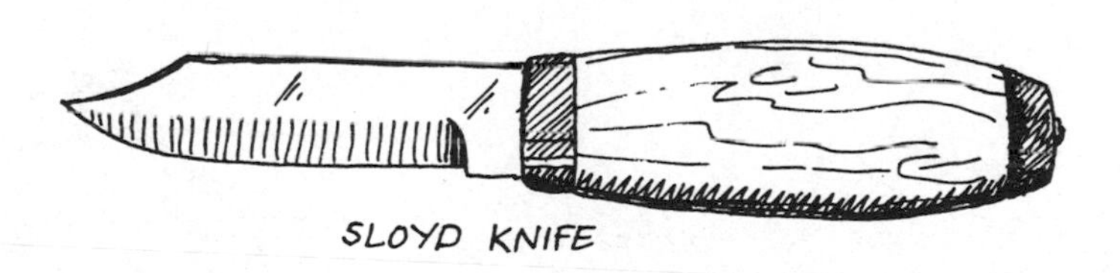

SLOYD KNIFE

Also used to make accurate lines is a **marking** or **scratch awl**.

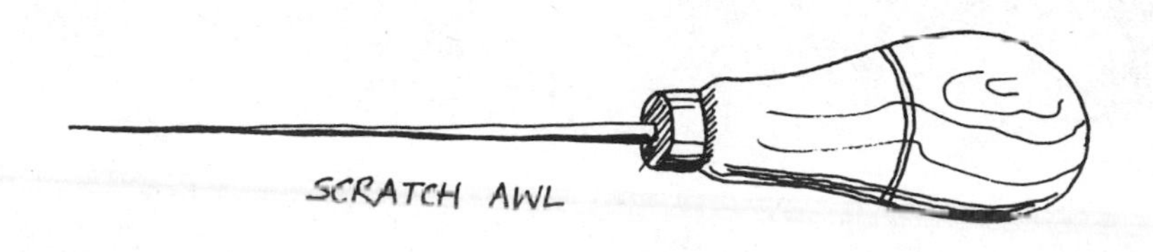

SCRATCH AWL

A different kind of marking tool is the **marking gauge**.

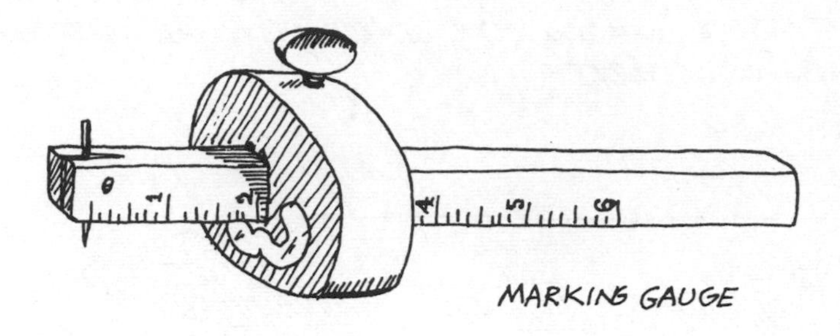

MARKING GAUGE

There are several different kinds of **marking gauges** for specific jobs, but basically they all mark lines parallel to the sides, the edges, or the ends of boards.

One other medium is used for marking wood, and that is chalk. The **chalk line** is one of the oldest ways of marking a long straight line. Old carpenters used to have to carry around a ball of string and a piece of chalk to rub on the string, which was stretched tightly between two points marking the end of the straight line they wanted to mark, and then, lifting the string a little, they would let it snap back onto the wood, where a line of chalk would remain after the string was removed. Nowadays, a **chalk line** is made like a large **tape measure**, the string and chalk being encased, keeping everything a lot cleaner.

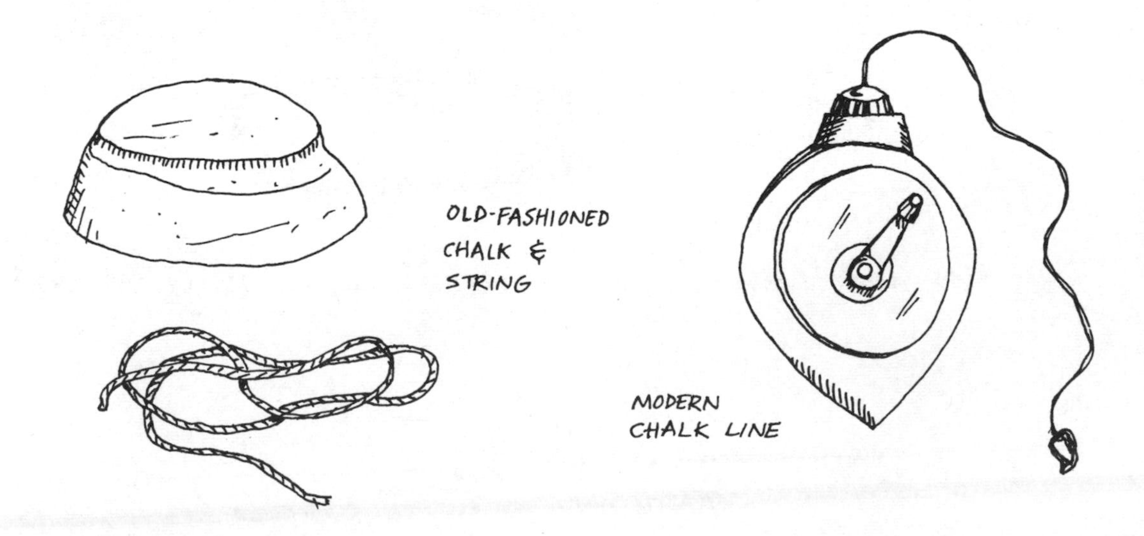

OLD-FASHIONED CHALK & STRING

MODERN CHALK LINE

Three last tools necessary for marking and measuring woodwork are the **try square**, **the bevel**, and the **framing square**.

The **try square** is a tool used for testing (and marking) lines perpendicular to an edge or a side of a board, and for seeing if edges and ends are square.

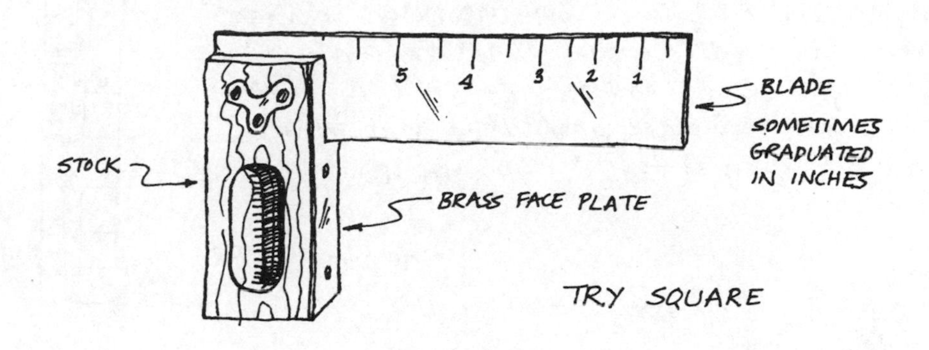

TRY SQUARE

There are whole sets of **try squares**, from tiny ones with 3" blades, to large ones with 18" blades.

A **bevel** is really an adjustable **try square**, for where the angle between the stock and the blade of the **try square** is always 90°, the **bevel's** blade can be adjusted to make any angle.

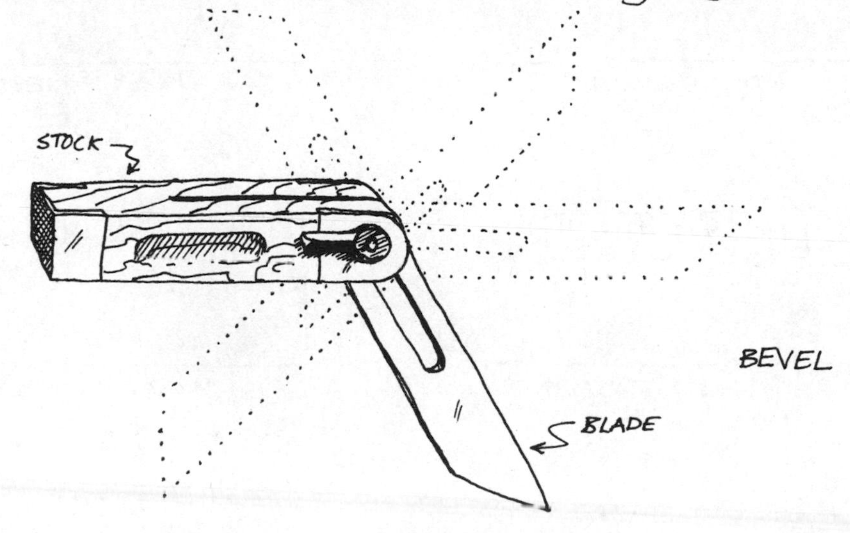

BEVEL

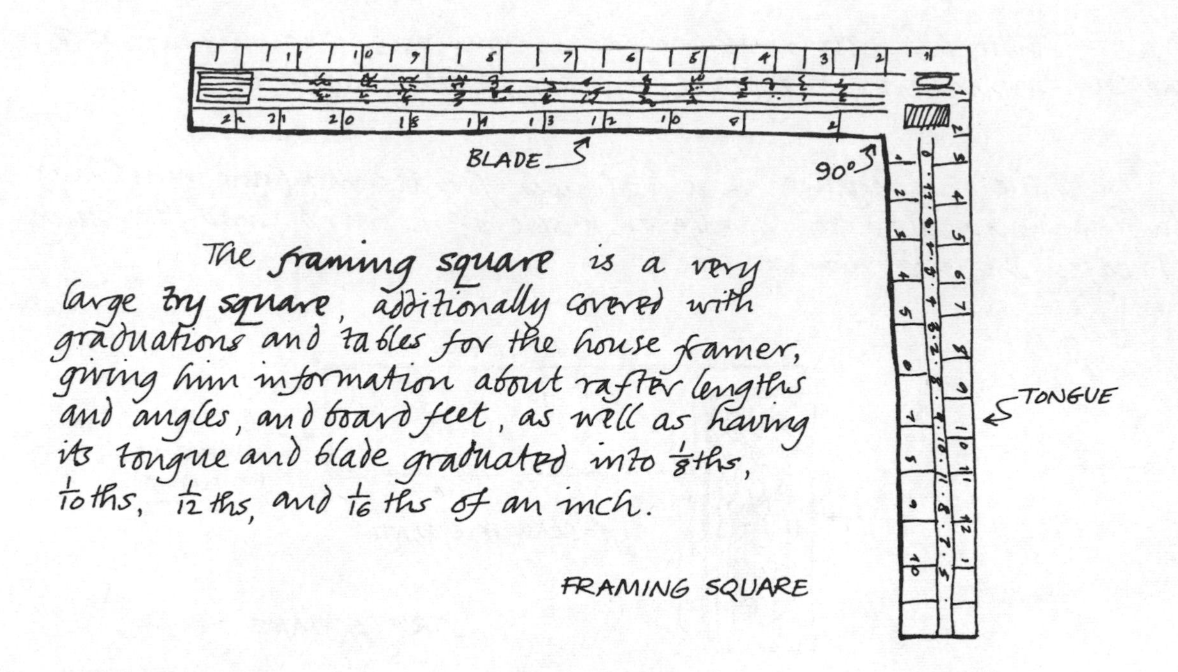

The **framing square** is a very large try square, additionally covered with graduations and tables for the house framer, giving him information about rafter lengths and angles, and board feet, as well as having its tongue and blade graduated into $\frac{1}{8}$ths, $\frac{1}{10}$ths, $\frac{1}{12}$ths, and $\frac{1}{16}$ths of an inch.

FRAMING SQUARE

USING MEASURING & MARKING TOOLS

Look at your rule, whichever kind you have, and see how it is marked. Notice that the graduation lines are varied in length to facilitate reading.

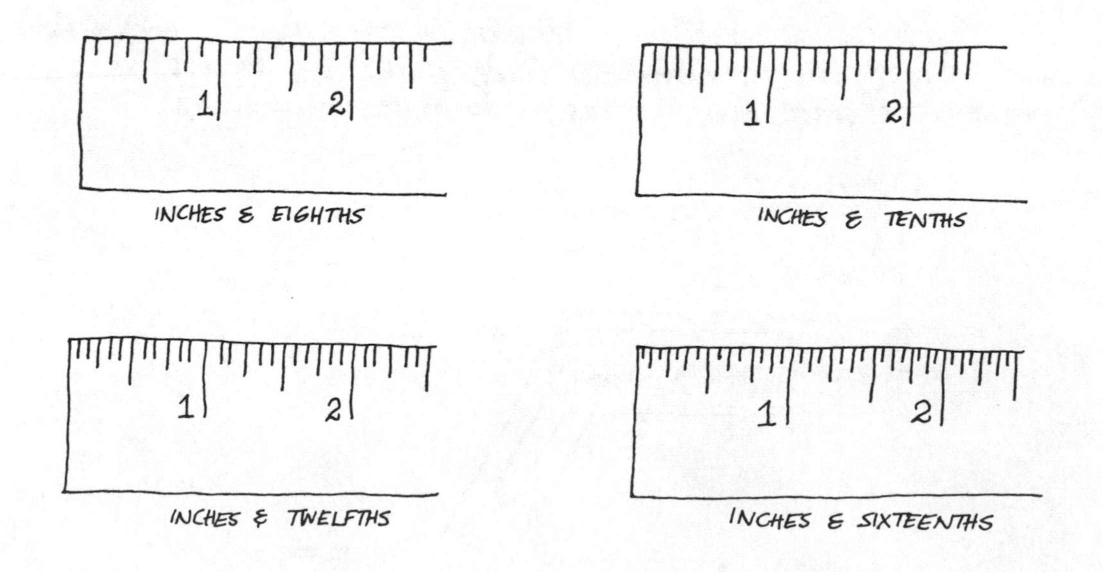

To measure a certain distance between two points with a **rule**, place the end of the **rule** even with one point and read the graduation line nearest the other point.

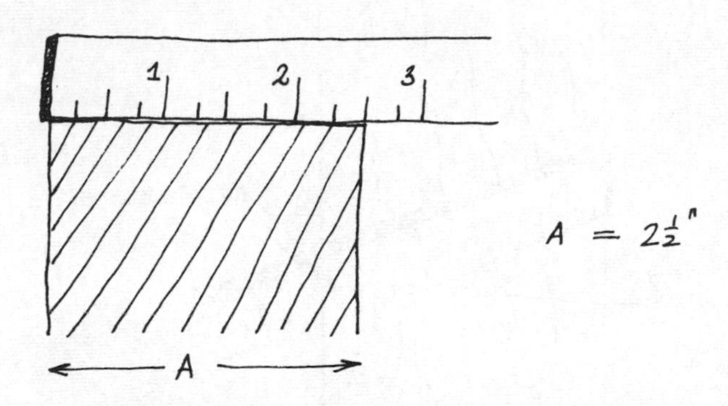

If the end of the **tape measure** or **rule** is worn or missing, start at the 1" mark and subtract 1" from your reading.

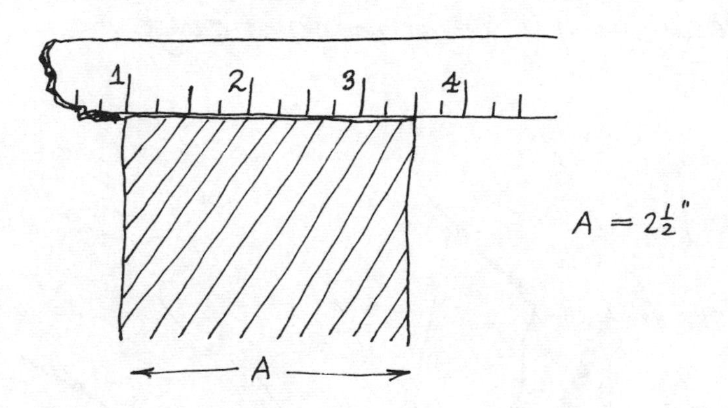

To measure accurately with a wooden **rule**, hold the **rule** on its edge so that the graduations touch the work being measured

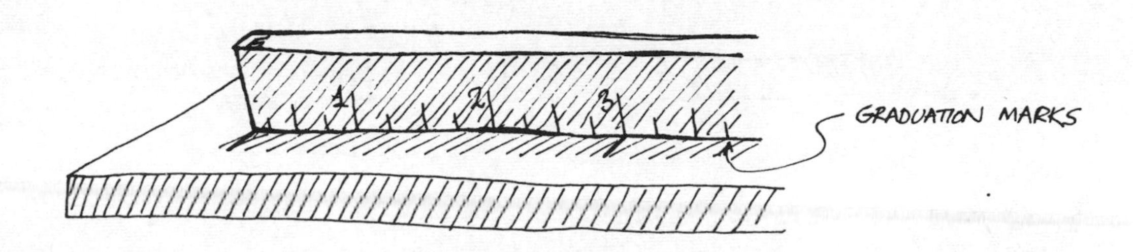

To find the middle of a board, hold the **rule** across the board so that two even inch marks (0 is even) touch the edges and then mark the mid-point.

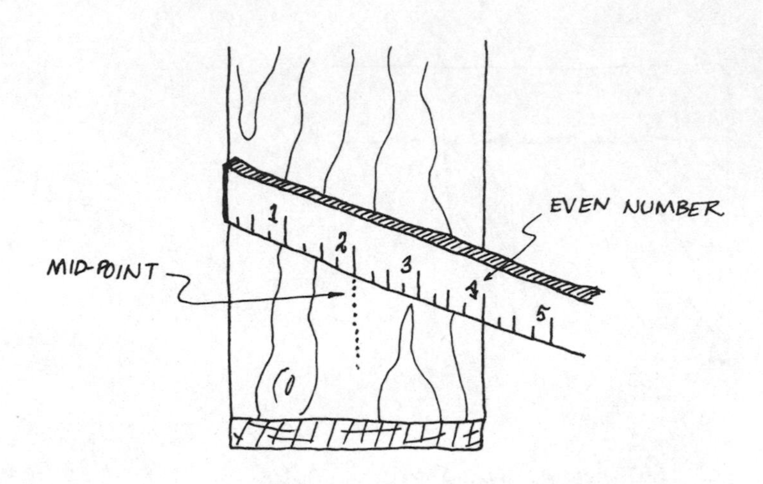

The **rule** can also be used to mark a continuous line parallel to an edge or side by using it like a **marking gauge**.*

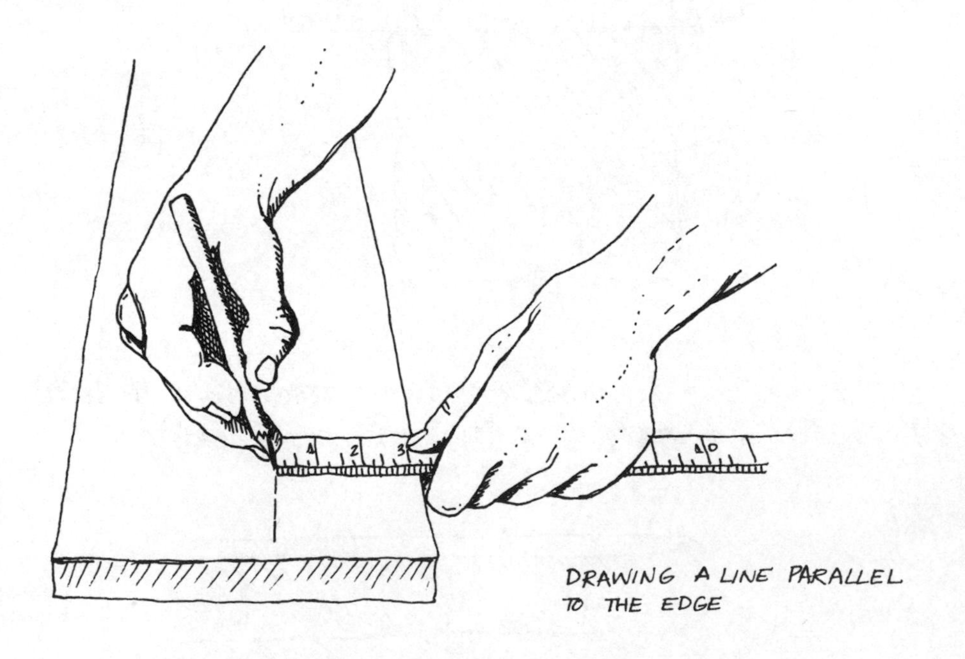
DRAWING A LINE PARALLEL TO THE EDGE

* see page 24

A **carpenter's pencil** should be used with the wide side against the **rule**.

If you rotate an **ordinary pencil** while drawing a line, the point will stay conical longer and need sharpening less often. Hold the **pencil** into the **rule** when drawing a line.

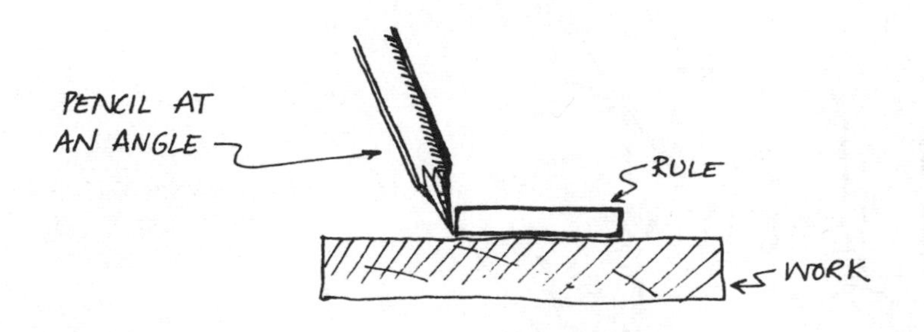

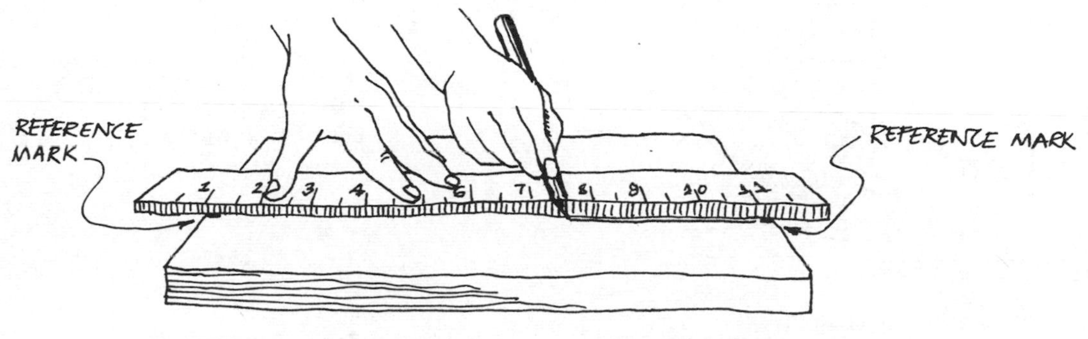

CONNECTING TWO POINTS WITH A STRAIGHT LINE BY MEANS OF REFERENCE MARKS

A **marking gauge** is used by first simply setting the head the required distance from the pin with a **rule**, and then drawing

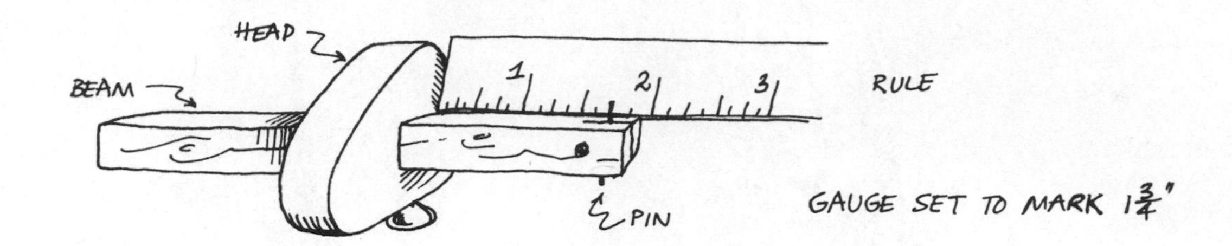

or pushing the **gauge** along the work with the head held closely to the edge, tilting the pin slightly so that it drags and does not dig in.

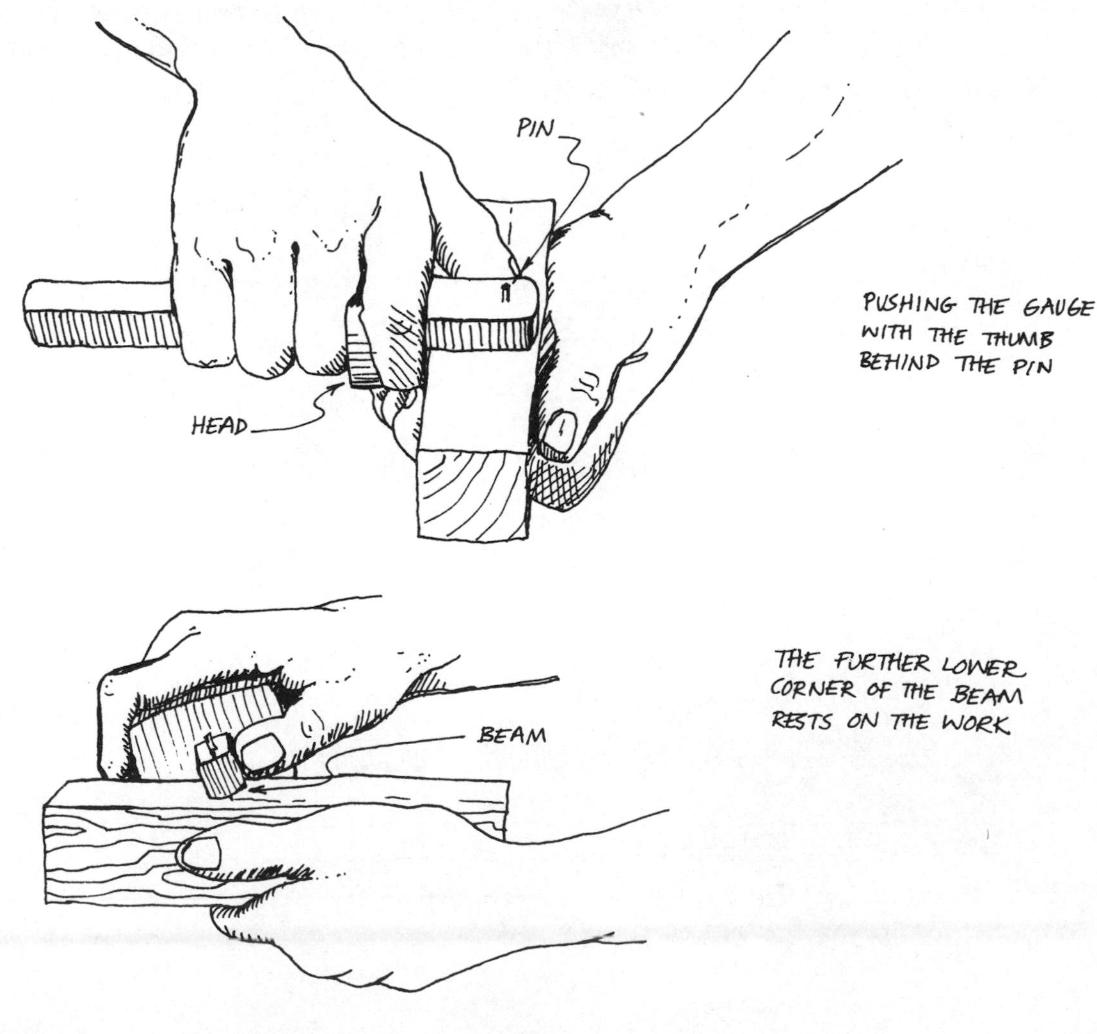

To stretch a **chalk line**, simply stretch it tautly between two points marking the ends of the desired line, either by tying one end to a nail or an **awl** stuck in the wood, or by holding it — sometimes it is easier to have someone else hold the far end than to stop and secure it with a nail — raise the line vertically and then let go.

POSITION OF THE HANDS WHEN SNAPPING A CHALK LINE

The **try square** is used to check squareness and flatness, but its most frequent use is for marking off the end of a board to be sawn.

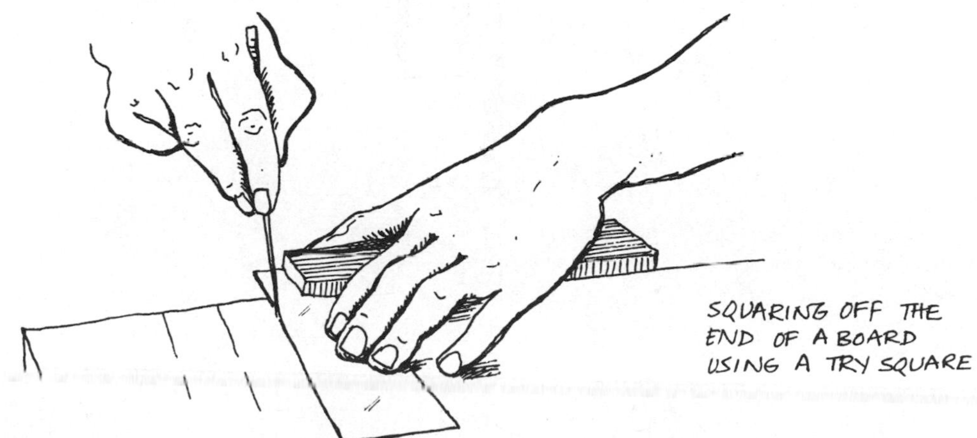

SQUARING OFF THE END OF A BOARD USING A TRY SQUARE

To test the surface of the work for flatness place the **try square** as shown, and see if the blade sits evenly or rocks.

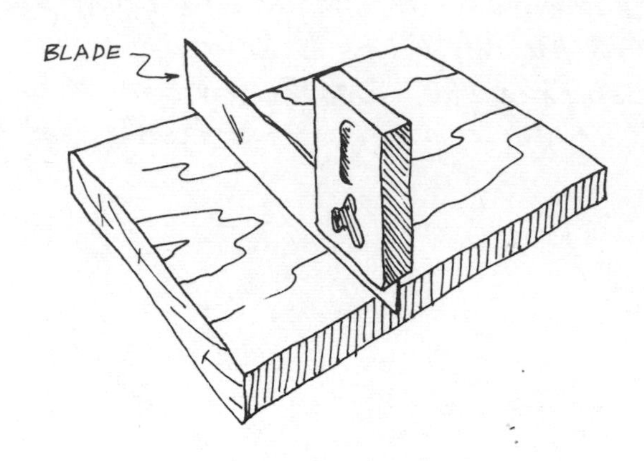

To test if the edge of a board is square, hold the handle of the **try square** against the surface of the board and move it up and down to see if the blade constantly touches the edge of the board.

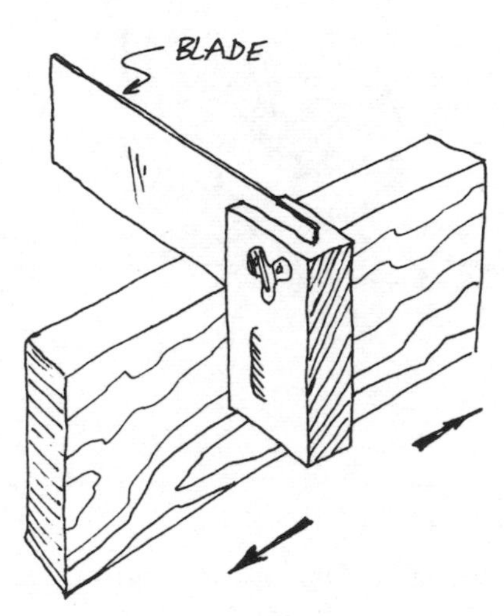

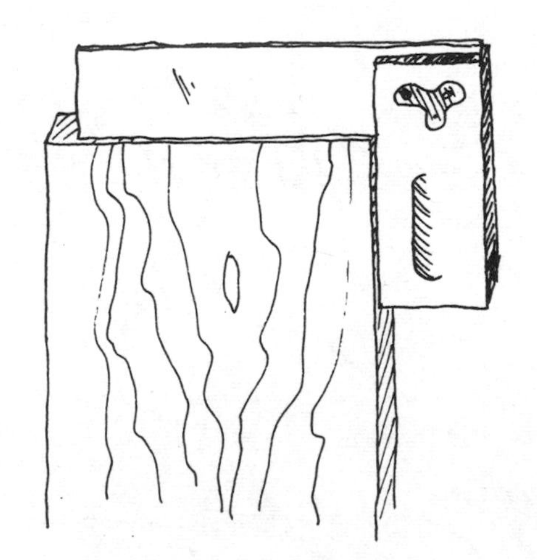

CHECKING IF THE END OF A BOARD IS SQUARE

A **bevel** is used for laying out and marking all angles other than 90°, for which the **try square** is used. The blade can be set to any desired angle either by using a **protractor** or by adjusting the blade to fit the angle of the work.

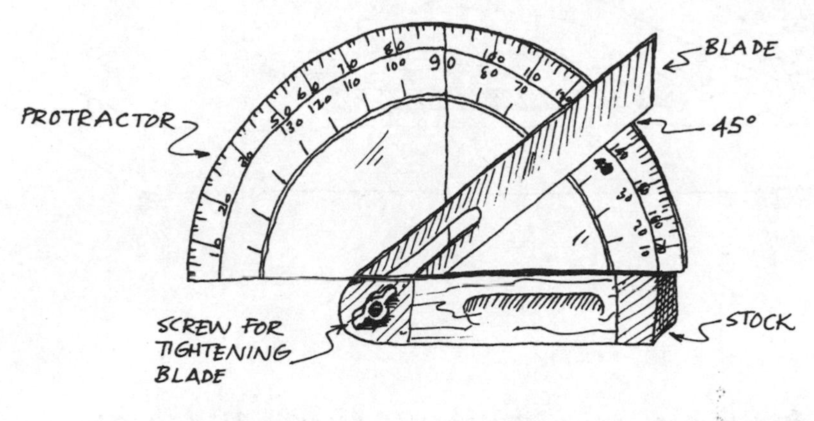

SETTING A BEVEL TO 45°

HOLDING TOOLS The basic holding tool, used by all those woodworkers who work at a bench, is the **bench vise**. There are many **vises**, but those used for wood are distinguished by having their jaws lined with wood, in order not to mar the work.

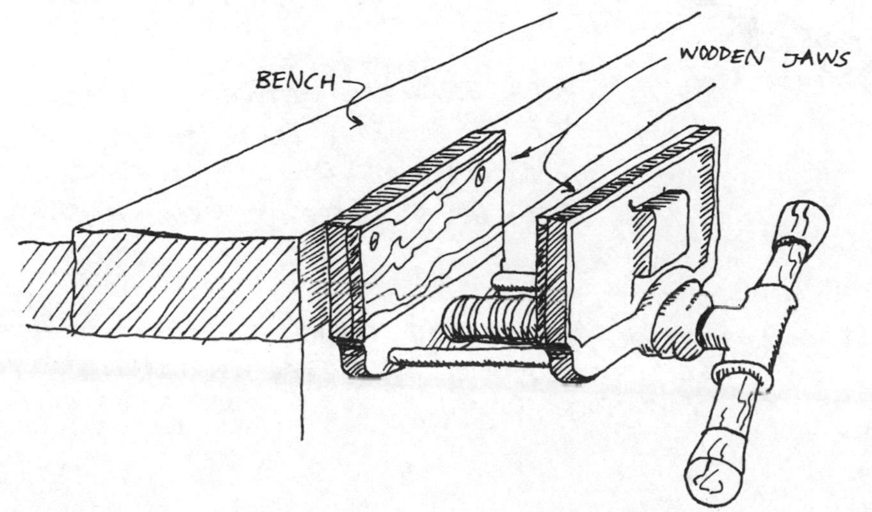

BENCH VISE

A **bench stop** can be a sophisticated built-in device on a workbench, or it can be something very simple, improvised on any handy work surface. Its function is to prevent the work from moving around by blocking its path. In its simplest form it consists of a small block of wood nailed to the bench or saw-horse.

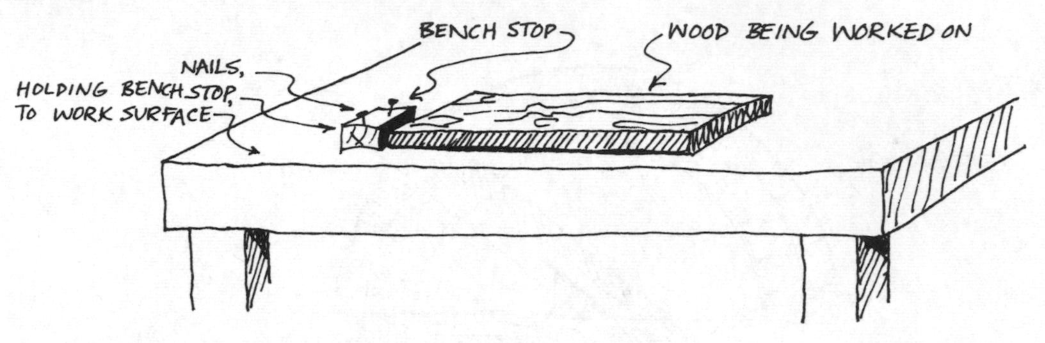

A **bench stop** can sometimes be used in conjunction with a **vise**.

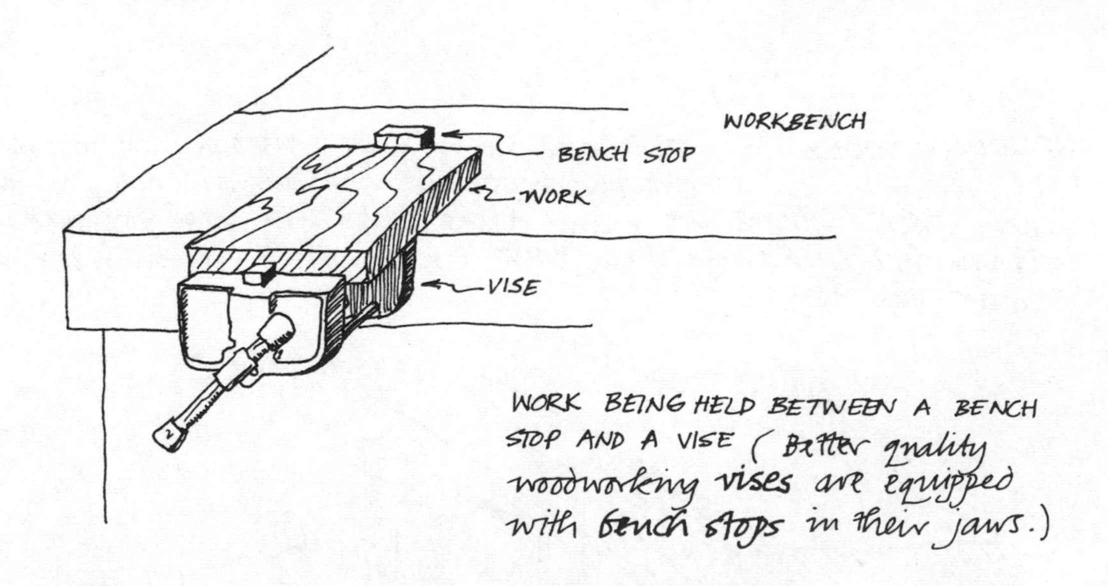

WORK BEING HELD BETWEEN A BENCH STOP AND A VISE (Better quality woodworking **vises** are equipped with **bench stops** in their jaws.)

SPRING

A factory-made **bench stop**, which is let into a hole drilled in the work surface, the spring holding it firmly at the required height.

For odd sawing and chiseling, so the bench will not be marred, and to have something to hold the work against, a **bench hook** is used. They may be bought, but they are such simple things that most woodworkers make their own out of scraps.

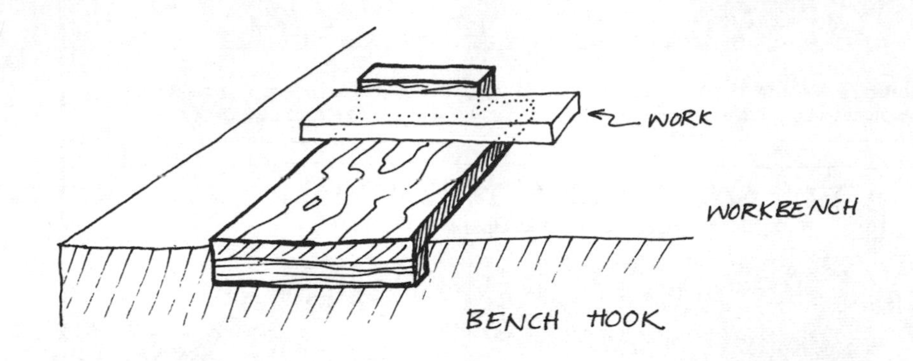

BENCH HOOK

When there are no **benches or vises** available you can hold things securely by means of various **clamps**, which can be used to hold the work down to sawhorses etc. The primary use of **clamps** is to hold wood tightly while being glued, and so they are made in a variety of shapes and sizes to accomodate all kinds of projects.

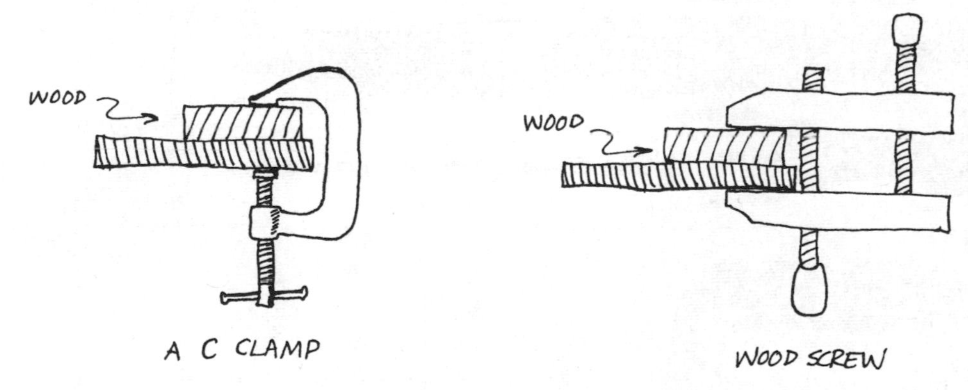

A C CLAMP

WOOD SCREW

In both the above **clamps** the bottom piece of wood could be the workbench.

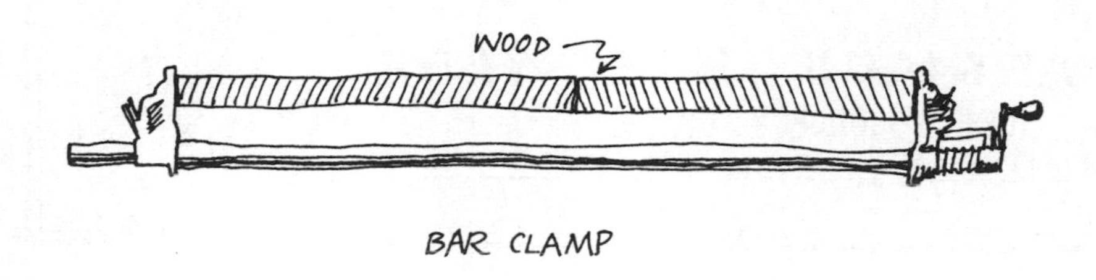

BAR CLAMP

Established 1854.

STEPHENS & CO.,

Manufacturers of U. S. STANDARD BOXWOOD and IVORY RULES.

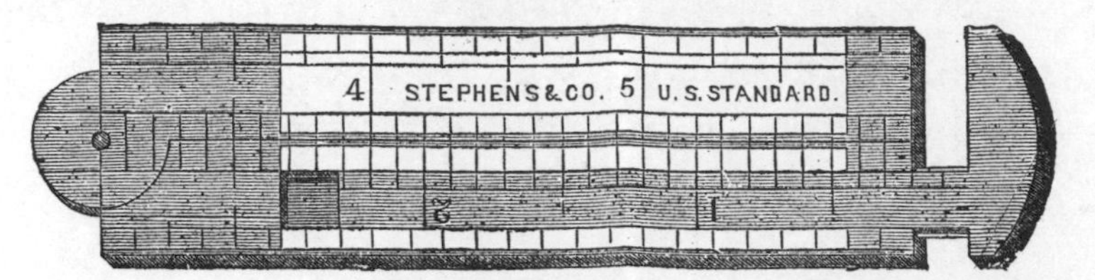

Also Exclusive Manufacturers of **L. C. Stephen's Patent Combination Rule.**

Rules graduated in foreign measure to order. *RIVERTON, CONN.*

E. M. BOYNTON,

Manufacturer of all kinds of

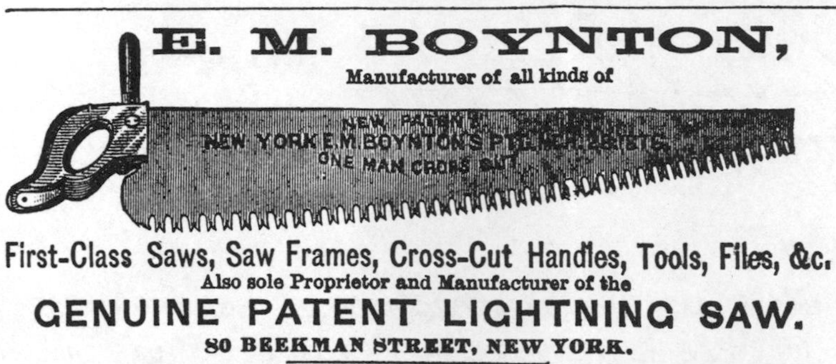

First-Class Saws, Saw Frames, Cross-Cut Handles, Tools, Files, &c.

Also sole Proprietor and Manufacturer of the

GENUINE PATENT LIGHTNING SAW.

80 BEEKMAN STREET, NEW YORK.

"BOYNTON'S SAWS were effectually tested before the judges at the Philadelphia Fair, July 6th and 7th. An ash log, 11 inches in diameter, was sawed off, with a 4½ foot lightning cross-cut, by two men, in precisely 6 seconds, as timed by the chairman of the Centennial Judges of Class Fifteen. The speed is unprecedented, and would cut a cord of wood in 4 minutes. The representatives of Russia, Austria, France, Italy, Spain, Belgium, Sweden, England, and several other countries, were present, and expressed their high appreciation." Received Medal and Highest Award of Centennial World's Fair, 1876. $1000 challenge was prominently displayed for six months, and the numerous saw manufacturers of the world dared not accept it, or test in a competition so hopeless.

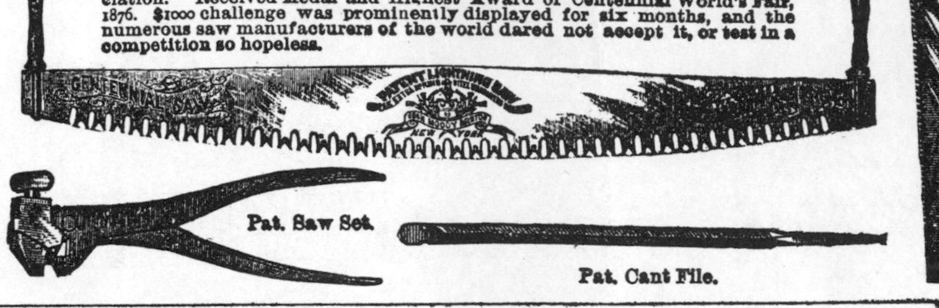

Pat. Saw Set.

Pat. Cant File.

Wood Workers' Clamps,

To open 2, 2½, 3, 4, 5, 6, 7, 8 and 10 inches.

Strongest and Best Clamp Made.

Malleable Ox Shoes

with

Steel Converted Toe Calk.

Five sizes.

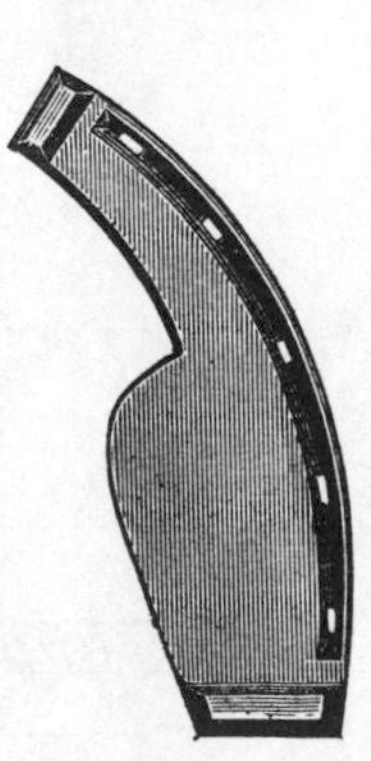

CARRIAGE HARDWARE,

IVES, WOODRUFF & CO., Manufacturers,

(Send for Catalogue.) *MT. CARMEL, CONN.*

CHAPTER TWO

Sawing ~

TAKE CARE NOT TO REMOVE ANY PART OF THE GUIDING LINE WITH THE SAW!

In almost every woodworking job, after the wood has been measured and marked, the first thing to happen is that the wood is sawn to size, and to the shape required for the job. The **saw** is therefore a very fundamental woodworking tool. Furthermore, sawing can be hard work. So it is important to understand which kind of **saw** to use for a particular job, and how to use it, in order to do things most easily.

SAW TYPES

There are lots of **saws** ranging from big ones designed for cutting down trees and sawing logs into boards, down to small **saws** made for little things like cutting out keyholes.

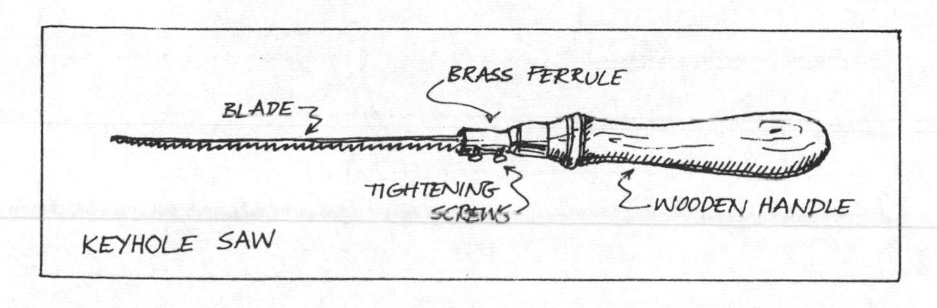

KEYHOLE SAW

Every type of **saw** has many variations but the basic types are as follows:

ONE-MAN and TWO-MAN CROSSCUT SAW

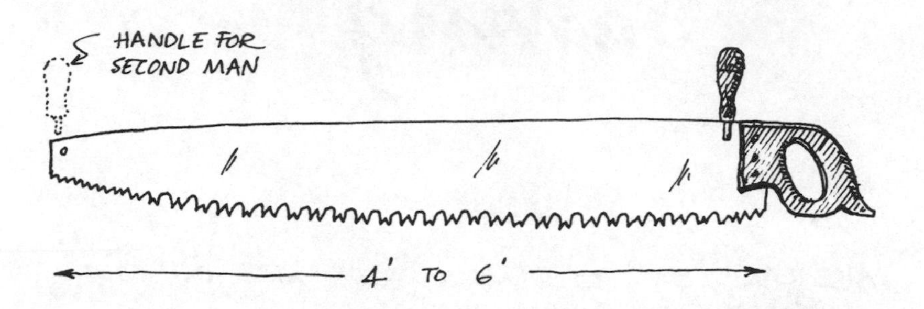

These long **saws** were designed for sawing trees and have largely been replaced by motor-driven **chain saws**.

FRAME SAW

A ***frame saw*** consists of a narrow blade fixed in a frame of wood stretched tight by a wire or thong at the opposite side of the frame from the blade. For centuries this was the commonest **saw**, and was made in all kinds of sizes for different jobs. It is rarely used today, except by cabinet makers, having been superseded by electric **table saws** and ***jig saws***.

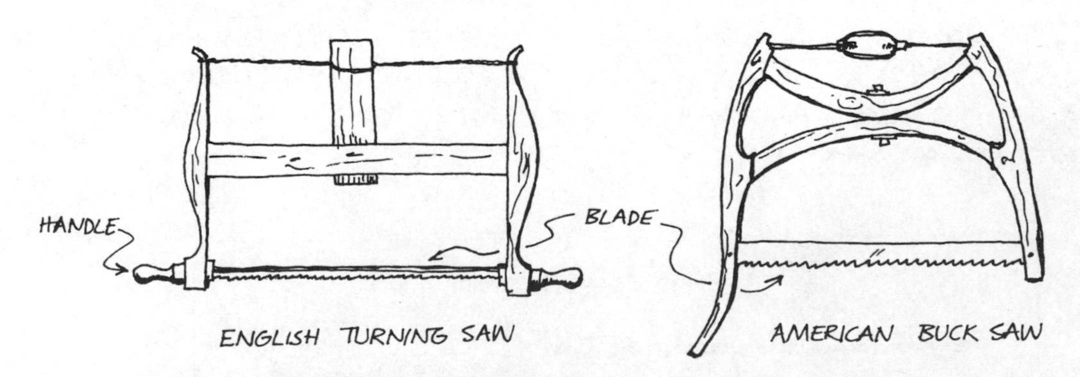

ENGLISH TURNING SAW

The blade may be turned, allowing the **saw** a deeper penetration when cutting curves.

AMERICAN BUCK SAW

This blade is fixed. The **saw** is used mainly for sawing firewood.

HAND SAW

The **hand saw** is today the commonest kind of **saw** used by all woodworkers. It consists of a thin, flat blade of steel with teeth along one edge and a wooden handle fastened to the large end by screws. The **hand saw** is made in more varieties than any other kind of **saw**. The two main types, however, are the **crosscut hand saw** and the **rip saw**, which will both be discussed in detail further on in this chapter. Basically, all **hand saws** look like this:

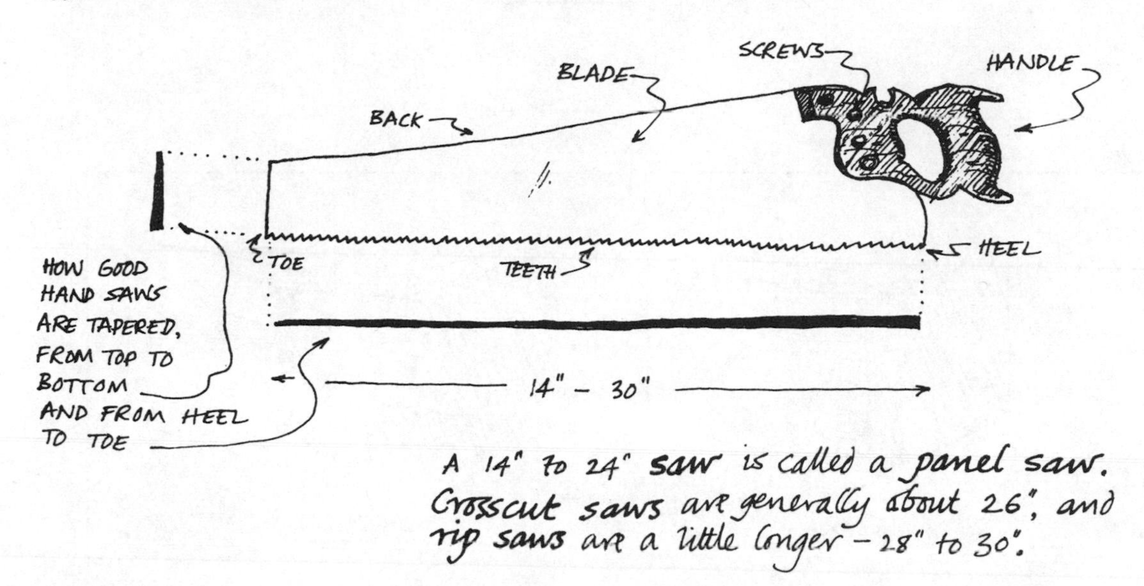

A 14" to 24" **saw** is called a **panel saw**. **Crosscut saws** are generally about 26", and **rip saws** are a little longer – 28" to 30".

BACK SAW

The **back saw** is a close-toothed **crosscut saw**, used in fine joinery and cabinet work, with a reinforced back, hence its American name. In Britain it is known as a **tenon saw** since one of its principle uses is to cut the tenon part of mortise and tenon joints – illustrated on the next page.

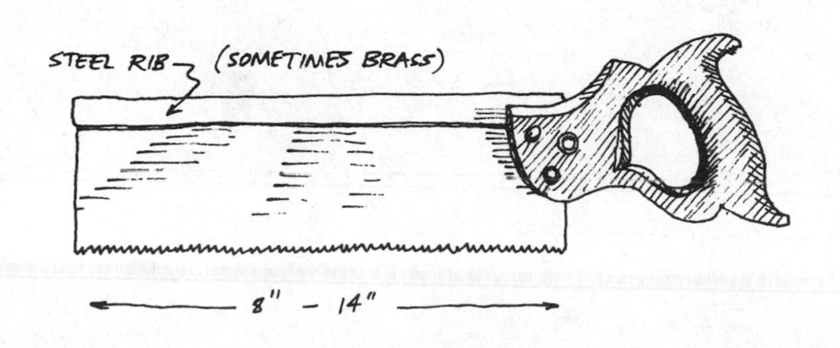

DOVETAIL SAW The **dovetail saw** is a smaller version of the **back saw**, used for even finer work, getting its name from its use in cutting dovetail joints.

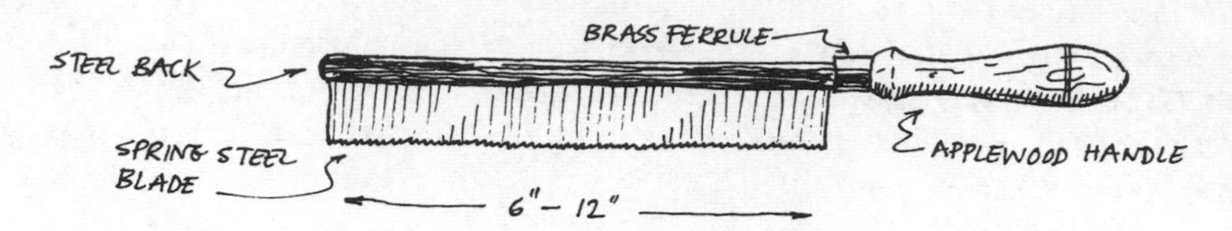

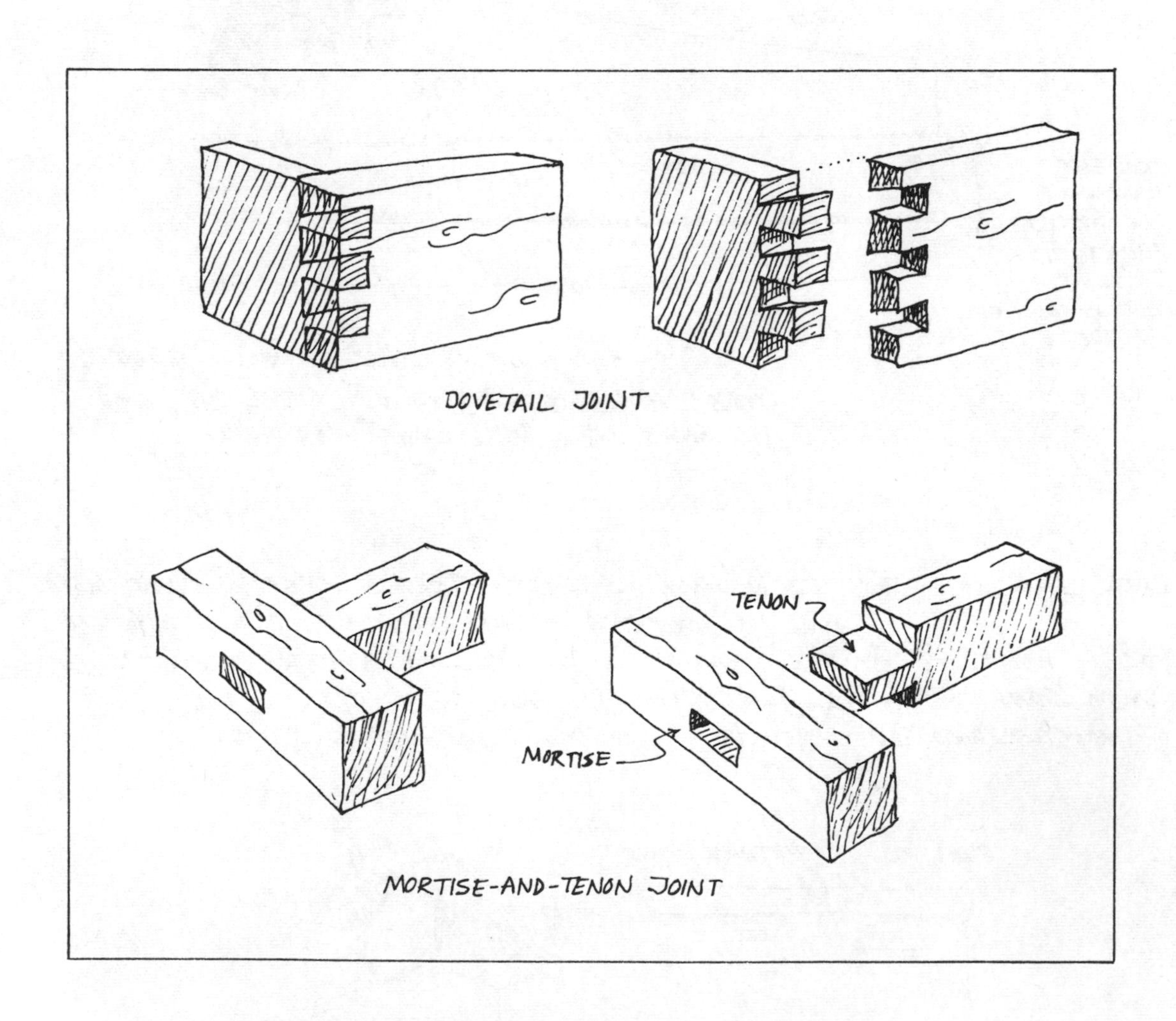

DOVETAIL JOINT

MORTISE-AND-TENON JOINT

COMPASS AND KEYHOLE SAW These **saws** are used for cutting curves in wood; consequently, the blades are very narrow, to allow them to turn.

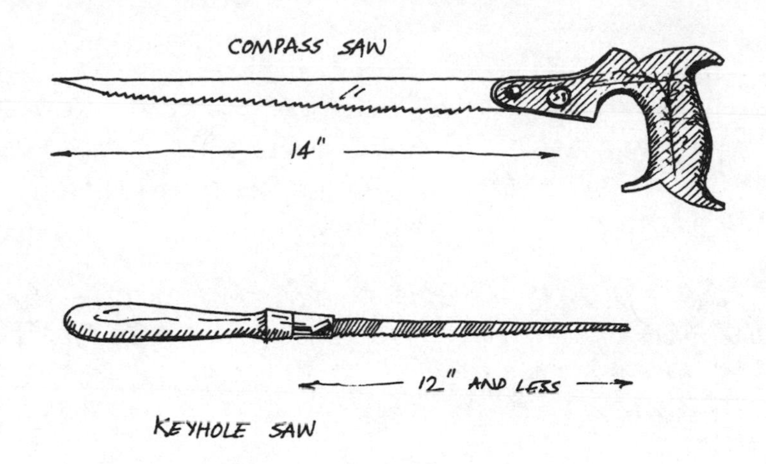

The **keyhole saw** is even thinner and shorter than the **compass saw**.

COPING SAW There are several varieties of **coping saw**, known as **fret saws** and **frame saws**. The difference lies in the shape of the frame which holds the blade, but they all have a very thin blade with teeth like a **rip saw*** in common.

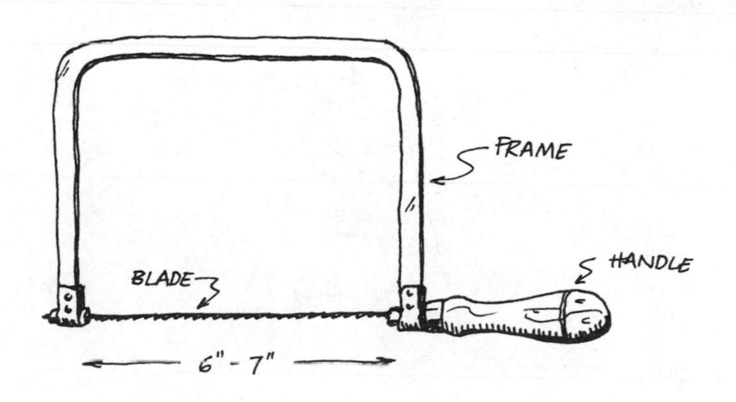

* rip saw, see page 39

CROSSCUT SAWING

The **crosscut handsaw** is used for cutting across the grain. It is distinguished from the **rip saw** by having more, and differently-shaped, teeth – generally 8 to 10 teeth per inch compared with the 5 to 7 teeth of the **rip saw**; see the illustration on page 39.

To use the **crosscut saw**, mark a line across a board near the end, using a **square** as illustrated on page 25. Hold the board with your left hand, with your thumb on the line. Hold the **saw** in your right hand with your first finger extended along the side of the handle to steady it – just like pointing.

Now, place the cutting edge of the **saw** so that the teeth in the middle of the blade just touch the right-hand edge of the line you drew. The blade should rest gently against your left-hand thumb, and make a right angle with the surface of the wood.

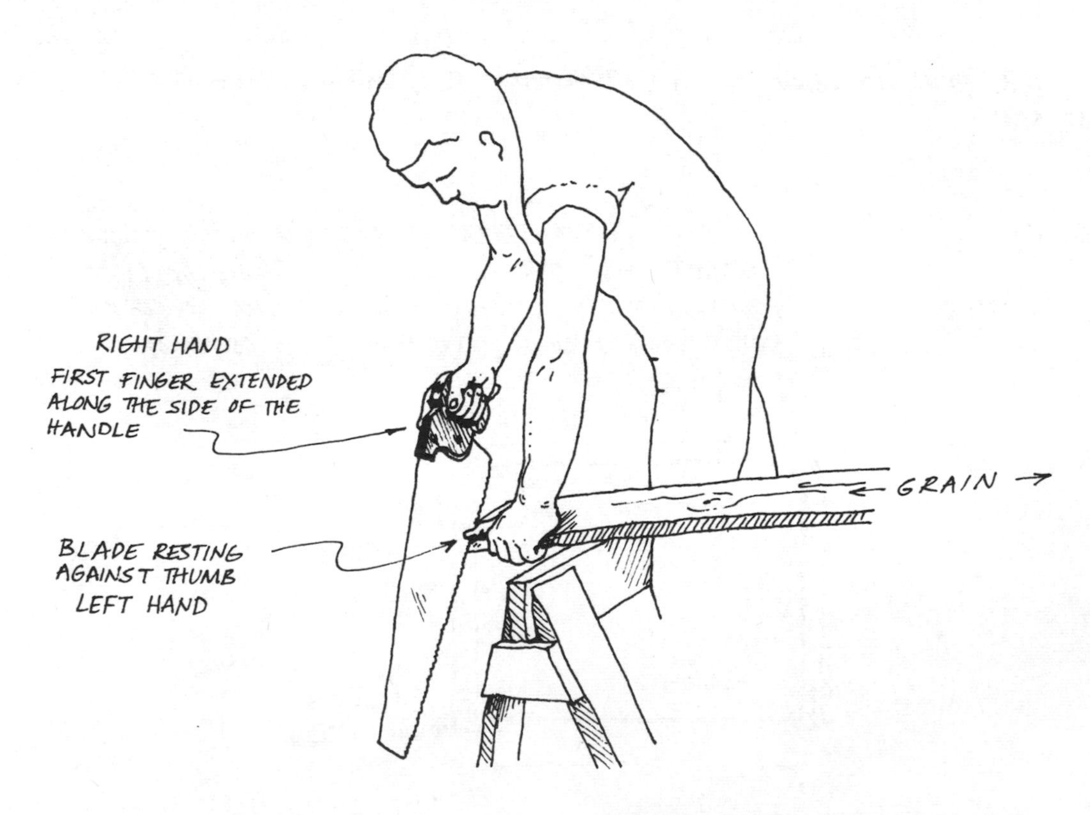

STARTING TO SAW

Without pressing on the **saw**, draw it several inches towards you in a straight line. This will start the cut, or as it is properly called, the kerf. With very slight pressure, reverse the motion just completed and move the **saw** back. Continue this gentle back and forth motion, increasing the length of the stroke each time, until the kerf is well started. This helps prevent the blade from moving sideways and tearing or splintering the edge of the board.

To guide the **saw** correctly, watch its path closely from directly above the line. If it leaves the line, twist the blade, while it is moving, in the kerf. Do not press on the **saw**, for then it will be more liable to leave the line, it will not cut so smoothly, and you may injure the blade. Also, it is more liable to jump out of the kerf and cut your hand if you are pushing hard. Let the **saw** do the work!

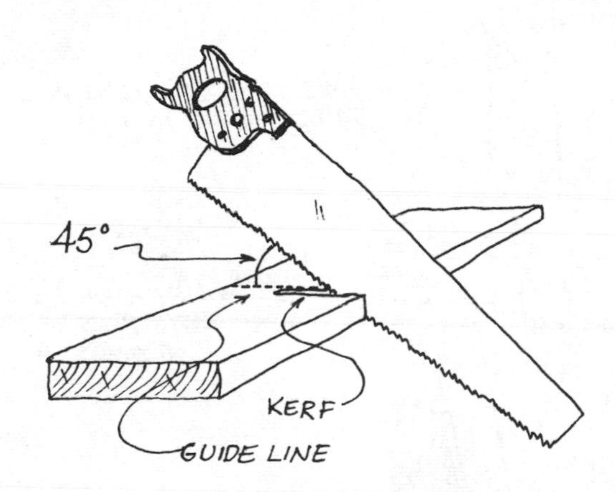

REMEMBER TO SAW TO THE RIGHT OF THE GUIDE LINE !

KEEP THE SAW AT AN ANGLE OF ABOUT 45° WITH THE WOOD.

TO CHECK THAT YOU ARE MAKING A SQUARE CUT HOLD A TRY SQUARE TO THE BLADE AND THE BOARD

When the board is nearly cut put your left knee against the board to prevent it from moving. Hold the board now on both sides of the kerf to support the piece being sawn off, otherwise its own weight will break it off before you have sawn all the way through and the edge will be splintered. If the piece being sawn off is too big to be held with one hand, provide some other support. Near the very end of the cut, take short, light, quick strokes, to prevent splintering the edge.

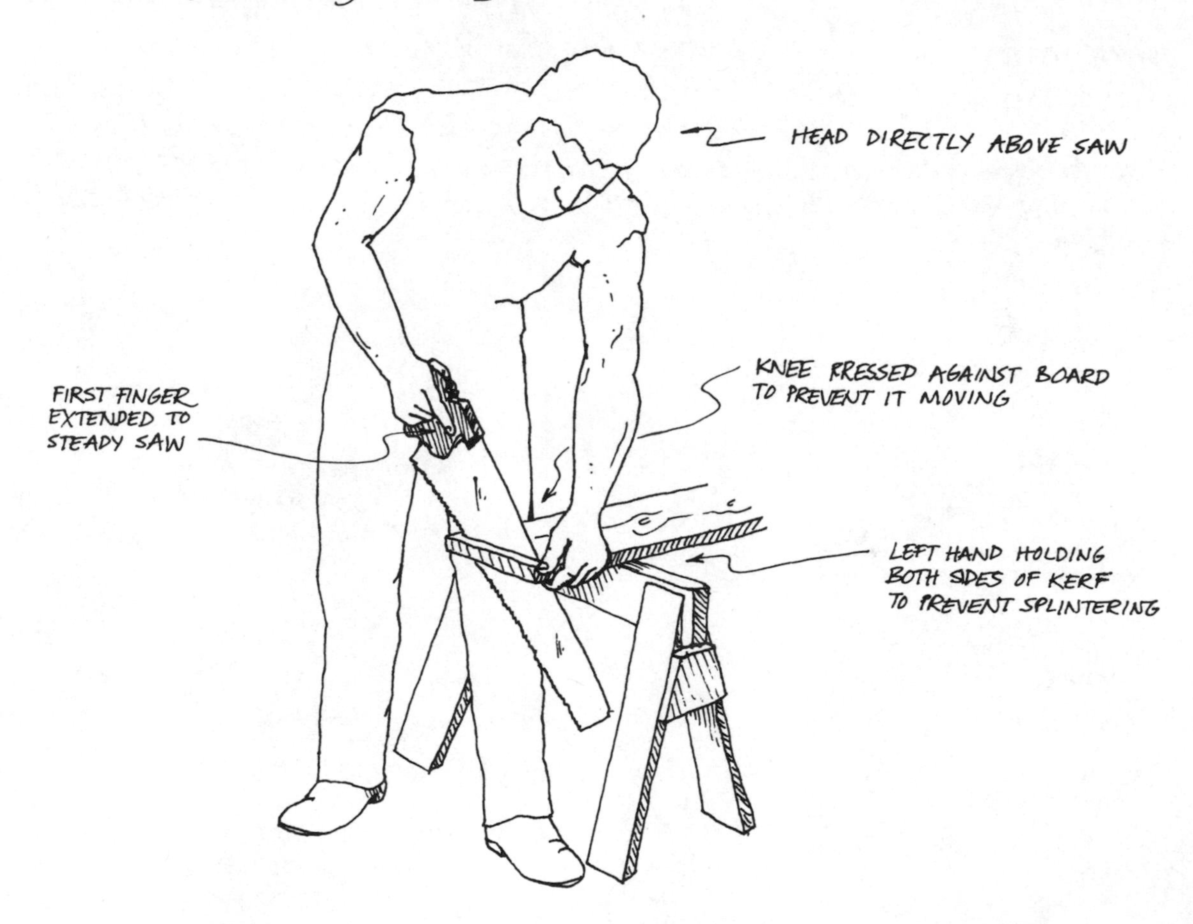

Keep the handle to the **saw** screwed on tightly. If it is loose and wobbly, you are wasting effort. Don't let your **saw** get rusty, it's harder to use that way; keep it clean with a thin film of oil.

RIP SAWING The **rip saw** looks very much like the **crosscut saw**, but it is used not for sawing across the grain like the **crosscut**, but for sawing with the grain. The difference between the **saws** lies in the way the teeth are made.

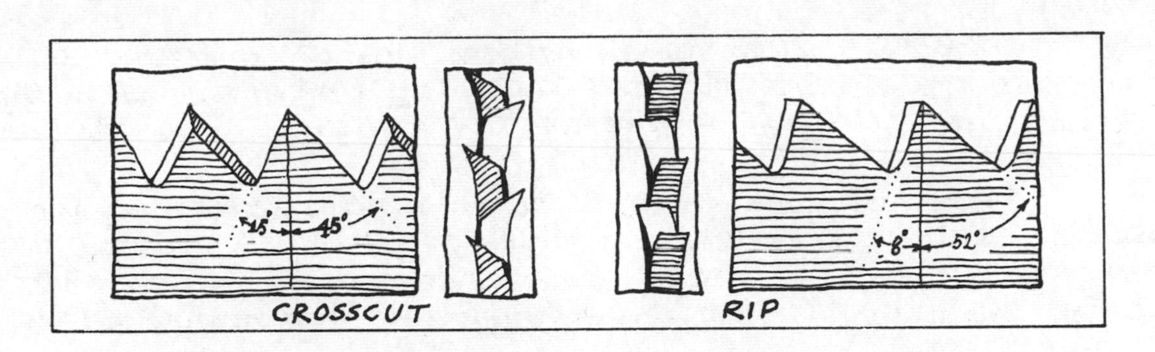

The **crosscut saw** has teeth like little knives, which slice through the wood. The **rip saw**, which has a heavier blade, has far fewer teeth per inch. These teeth are bigger and made at a different angle, being more like little **chisels** – each tooth cutting away a tiny chip of wood.

The reason for the difference is that the differently shaped teeth make it easier to cut across the grain with a **crosscut saw** than with a **rip saw**, and easier to cut with the grain with a **rip saw** than with a **crosscut saw**.

Try using a **rip saw** to cut across the grain, and notice how it tears the wood. Then try using a **crosscut saw** with the grain, and notice how the **saw** "binds", or gets stuck in the wood.

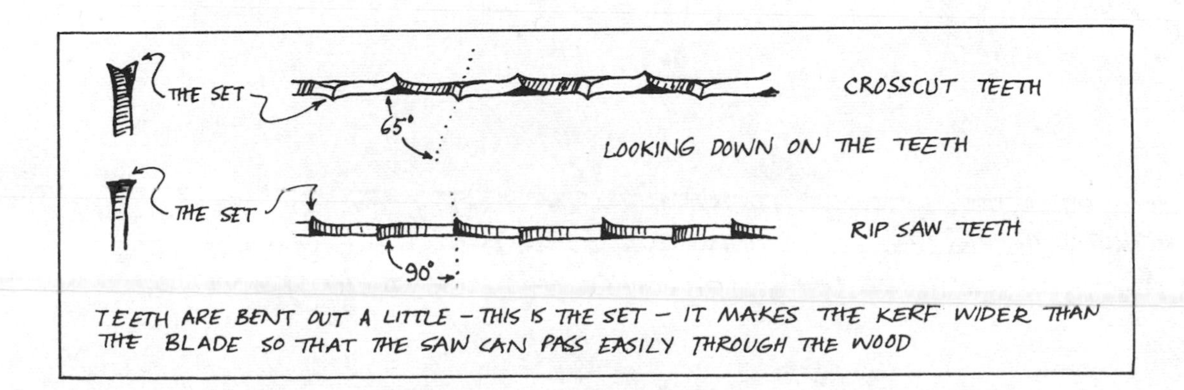

When using the **rip saw** to saw along a board, that is, <u>with</u> the grain, begin as with crosscut sawing, taking short strokes until the kerf is well established, and then lengthening the stroke until all the blade is being used. Saw close to the line, but remember, take care not to remove any part of the guide line with the **saw**.

Generally, when rip sawing from the butt, or bottom, of a tree towards the top, the parts spread; and when sawing in the opposite direction, the parts come together and tend to bind the **saw**.

It is not always easy to tell in which direction you are sawing, so to prevent the **saw** binding, you must insert a wedge in the kerf as soon as you have sawn a short way. Most carpenters use a **screwdriver**, although anything similar will do.

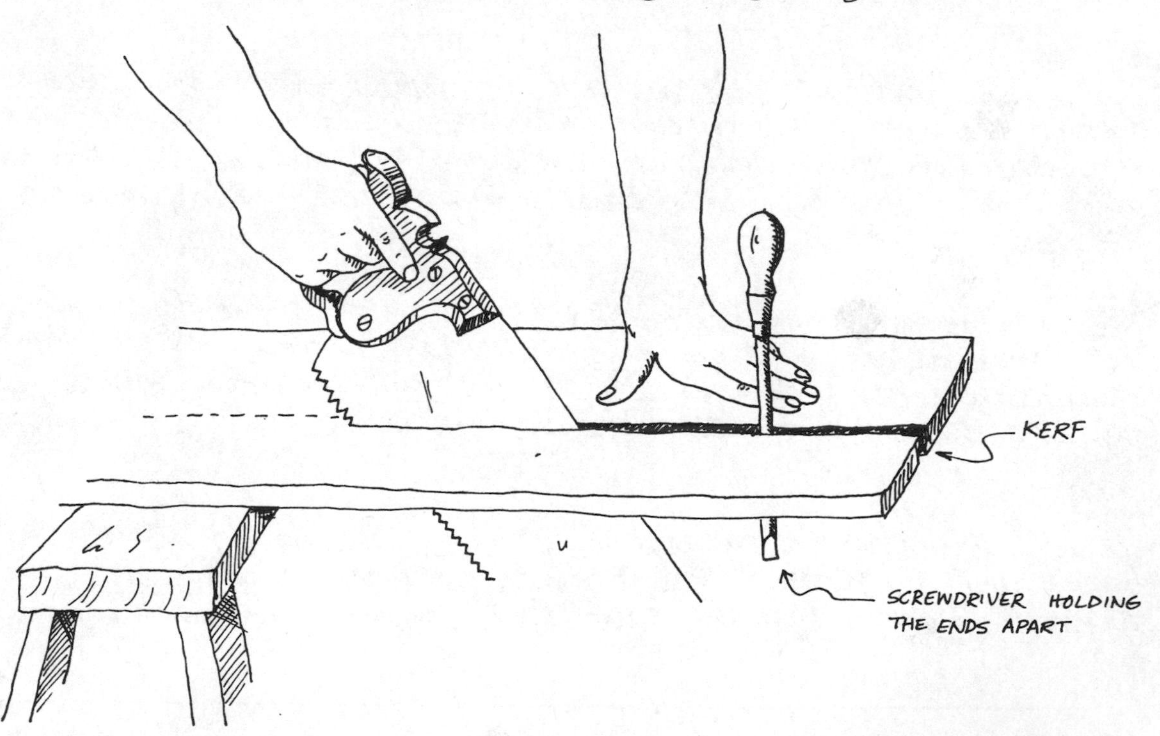

Keep your eye directly above the blade, taking care to hold the **saw** as straight as possible. Should the **saw** begin to wander from the line, correct it by twisting the handle until you are back to the guide line.

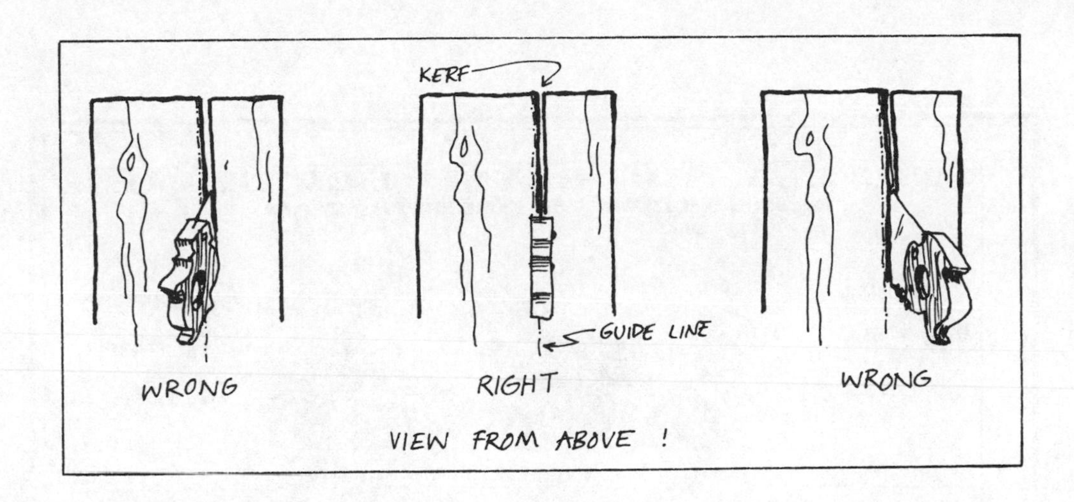

USING THE BACKSAW Unlike a **hand saw**, the **backsaw** is usually held flat on the wood when it is used, and not at an angle, although a slight angle for the first few strokes helps get the kerf started. The blade can be guided by the thumb of the free hand. When working on small pieces of wood, the work is generally held in a **bench hook**; see page 29.

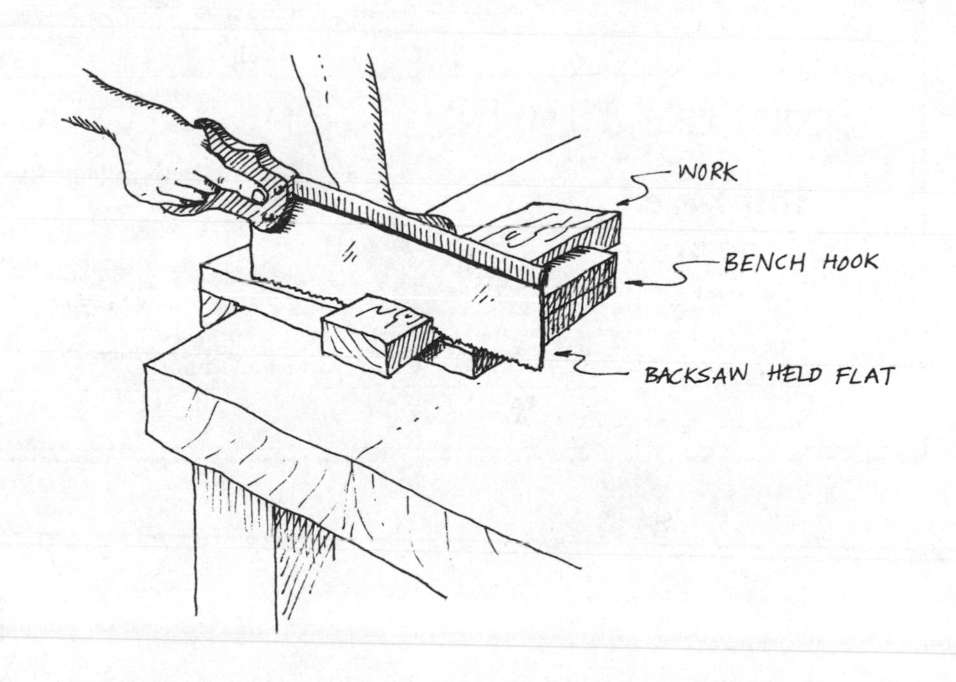

STANLEY RULE AND LEVEL CO.,

Manufacturers of **IMPROVED CARPENTERS' TOOLS.**

Factories,
New Britain,
CONN.

Warerooms,
35 Chambers
St., N. Y.

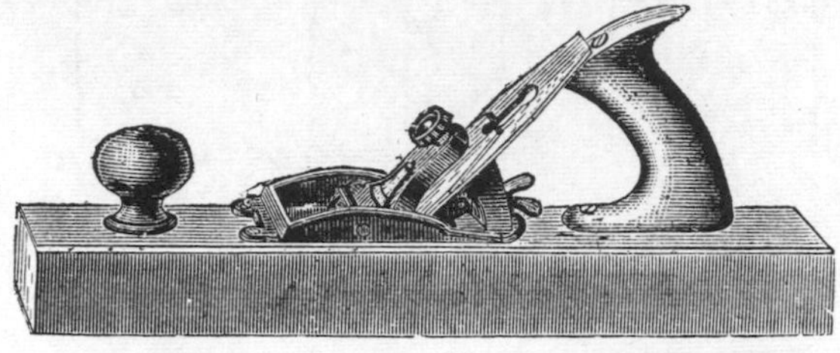

No. 129, Fore Plane, 20 inches in length. 2⅜ inch Cutter. $2·25.

LEONARD BAILEY & CO.,

MANUFACTURERS OF

STANDARD MECHANICS' TOOLS.

FACTORY, HARTFORD, CONN.

Sargent & Co., Agents.

WAREROOMS,

37 Chambers St., New York.

Trade Mark.	Pat. Jan. 4, 1876.
Victor Planes.	Pat. Dec. 12, 1876. " April 10, 1877.
Circular Plane.	Pat. March 28, 1871.
Dado Plane.	Pat. July 6, 1875.
Try Squares.	Pat. May 9, 1871. " Dec. 23, 1873. " Dec. 8, 1874.
Flush T. Bevel.	Pat. March 19, 1872.
Spoke Shaves.	Pat. Oct. 3, 1876.
Box Scraper.	Pat. Oct. 3, 1876.

The VICTOR PLANES are the product of twenty-two years' experience in manufacturing Iron Planes. They are simple, durable, and in every essential practical, cheap as the cheapest, and *warranted the best.*

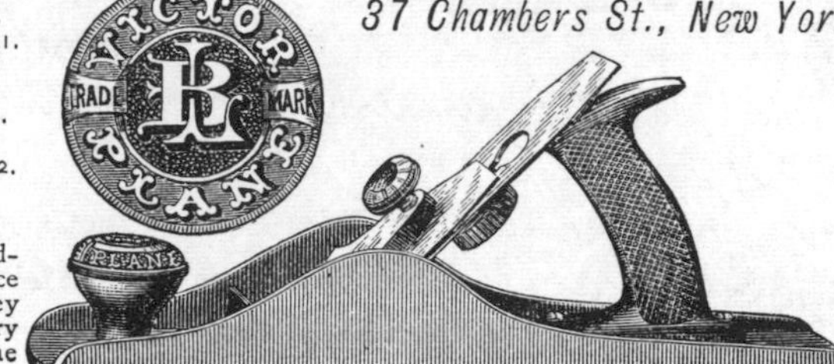

Send for Illustrated Catalogue and Price List.

Stanley Rule & Level Co.,

MANUFACTURERS OF

Improved Carpenters' Tools.

FACTORIES,
New Britain, Conn.

WAREROOMS,
29 Chambers St.,
New York.

Manufacturers of **Bailey's Patent Adjustable Planes.**
General Agents for the sale of **Leonard Bailey & Co.'s "Victor Planes."**
Manufacturers of **"Defiance" Patent Adjustable Planes.**

CHAPTER THREE

Planing ~

ALWAYS PLANE WITH THE GRAIN !

After wood has been sawn to the shape and size required for a job, the next step is very often to work it with a **plane**, which operation is called planing. A **plane** is a sharp-edged tool used to dress the wood to finished dimensions. To dress means, in woodworking, to make smooth, or to finish. Wood may be bought, for example, directly from a saw mill, with the marks of the **saw** still upon it, when it is called rough-sawn lumber, or planed and smoothed from a lumberyard, when it is referred to as dressed lumber.

Like **saws**, there are many kinds of **planes**, some of which are designed for extremely specialised jobs. Originally, however, **planes** were simply chisel-like blades set in blocks of wood, but they have been refined and developed into much more complicated tools.

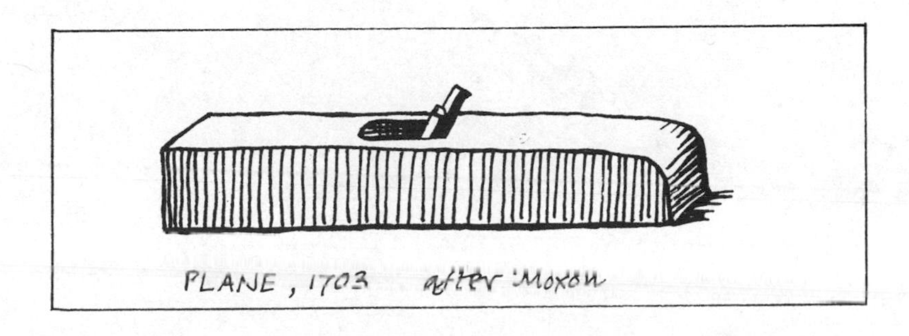

PLANE , 1703 after Moxon

PLANE TYPES

There are three classes of **planes**: **bench planes**, **block planes**, and **special-purpose planes**.

BENCH PLANES

These are the commonest **planes**, so called because they are most often used at a bench. There are lots of them, classified according to length, each length having a somewhat different function. The four main sizes are as follows:

The **smooth plane** is used for finishing off short or uneven surfaces.

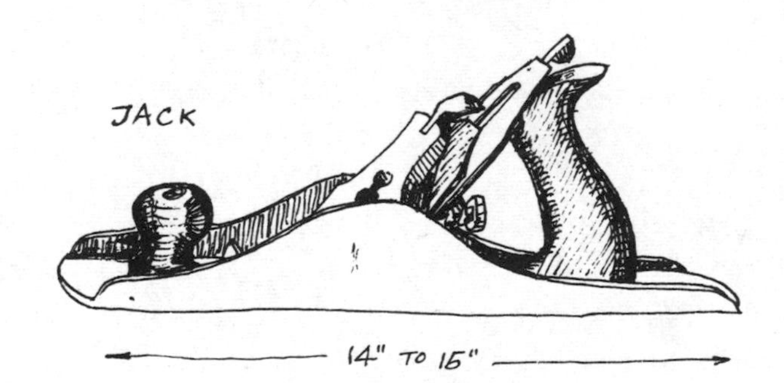

The **jack plane** is the jack of all trades as far as **planes** go. It can be used for rough work or fine work, sometimes replacing the **smooth plane** and sometimes being used instead of the **fore plane**, which is intended for long surfaces.

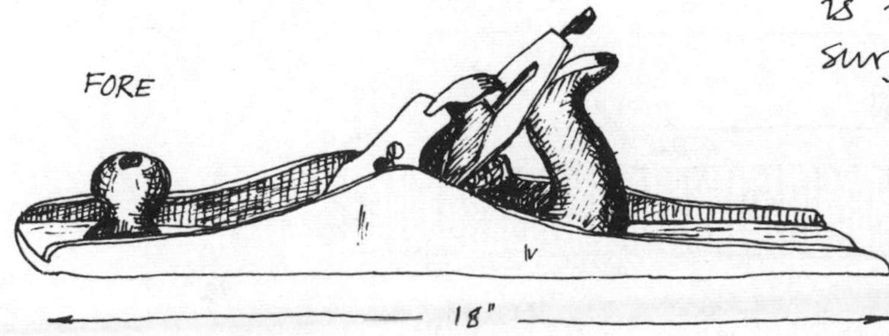

JOINTER

The **jointer plane** is the longest of the **bench planes**. With it the flattest surfaces may be obtained, and so it is used to true the edges of boards which must fit together.

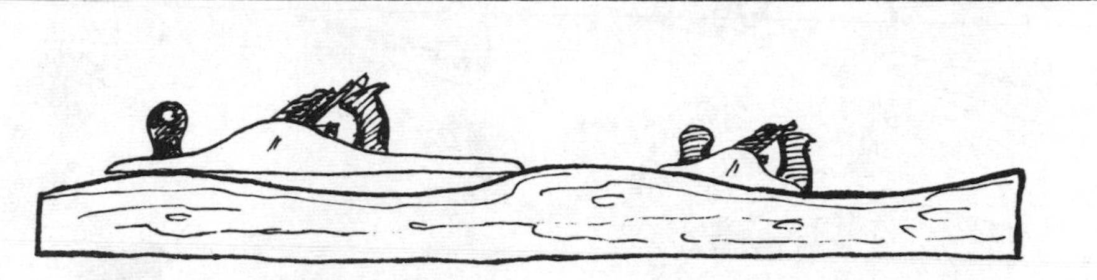

A LONG PLANE BRIDGES THE LOW PARTS AND CANNOT CUT THEM UNTIL THE HIGH SPOTS ARE REMOVED; A SHORT PLANE FOLLOWS THE CURVES.

BLOCK PLANE

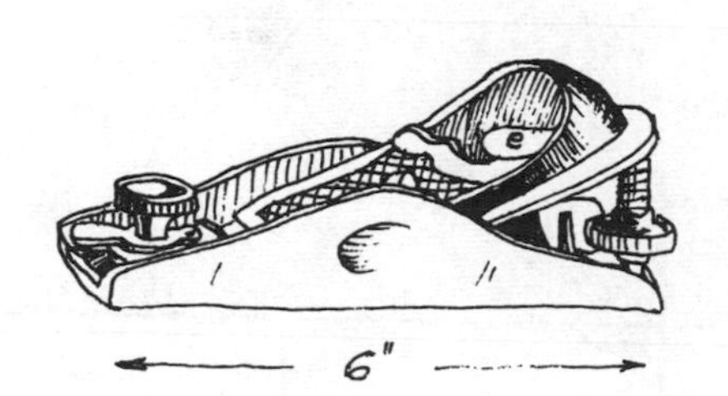

The **block plane** is a small one-handed **plane** with a single low-angle blade, which makes it ideal for planing end grain.

SPECIAL PURPOSE PLANES This is a very large group of extremely varied tools, including **planes** designed to cut grooves and slots of all descriptions, such as **rabbet planes, dado planes**, and **plough planes**; **planes** to be used in special places, such as very small **planes**, one-sided **planes**, or **planes** with no front to them; and **moulding planes**, which were once used a lot to make the intricate mouldings that used to adorn all kinds of woodwork.

RABBET PLANE

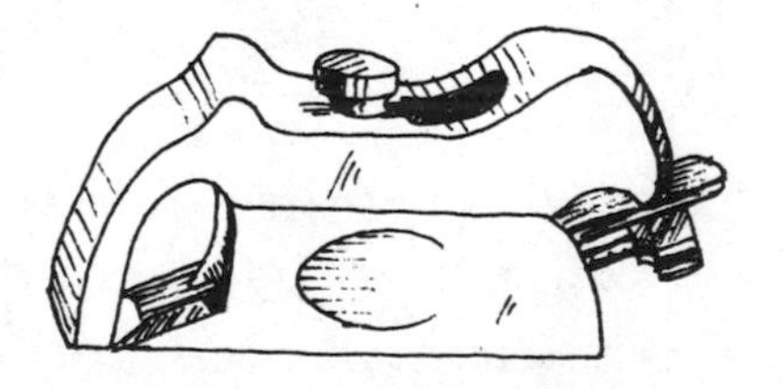

These **planes** are for fine cabinet work where extreme accuracy is required.

MOULDING PLANE

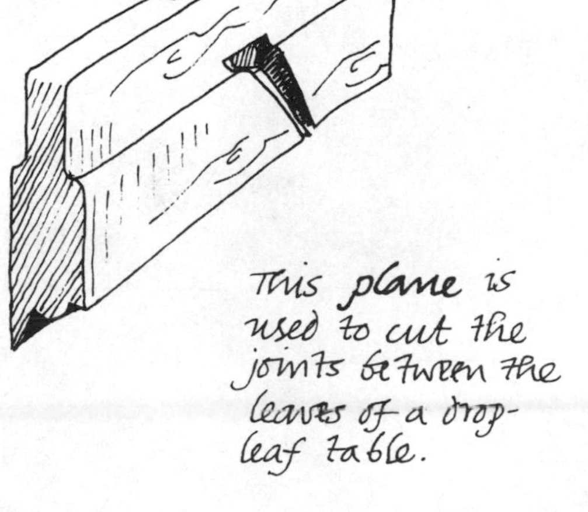

This **plane** is used to cut the joints between the leaves of a drop-leaf table.

PARTS OF A PLANE

The modern **iron plane**, which is an American invention, developed during the last century in order to "...simplify the manufacturing of **planes**: second to render them more durable; third to retain a uniform mouth; fourth to obviate their clogging; and fifth the retention of the essential part of the **plane** when the stock is worn out," * consists of a metal stock in which is set an adjustable double iron. Previous **planes** had wooden stocks and most commonly only single irons. The stock is the body, and the iron is the name for the blade.

19TH CENTURY WOODEN JACK PLANE

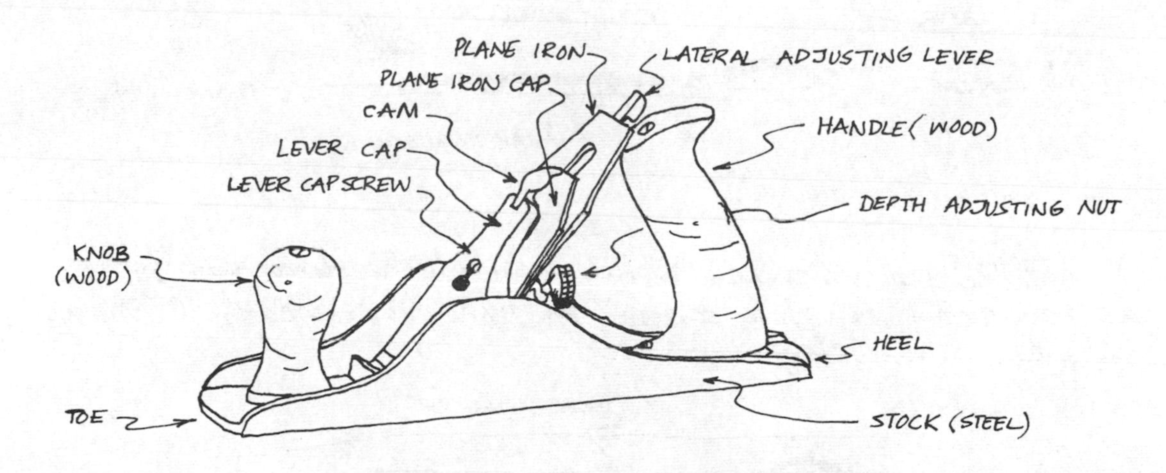

MODERN STANDARD IRON JACK PLANE

* M.B. Tidey, 1857 Patent applicant.

ADJUSTING A PLANE

Before you can use a plane successfully, there are four important adjustments which you must understand and know how to make. They concern:

1. The double plane iron.
2. The lever cap.
3. The depth of the cut.
4. The lateral setting of the iron.

1. THE DOUBLE PLANE IRON

The modern plane blade is made up of two parts: the plane iron, which has the cutting edge; and the plane iron cap, whose purpose is to break and roll the shaving cut by the plane iron so that a smooth surface is produced and the plane does not clog.

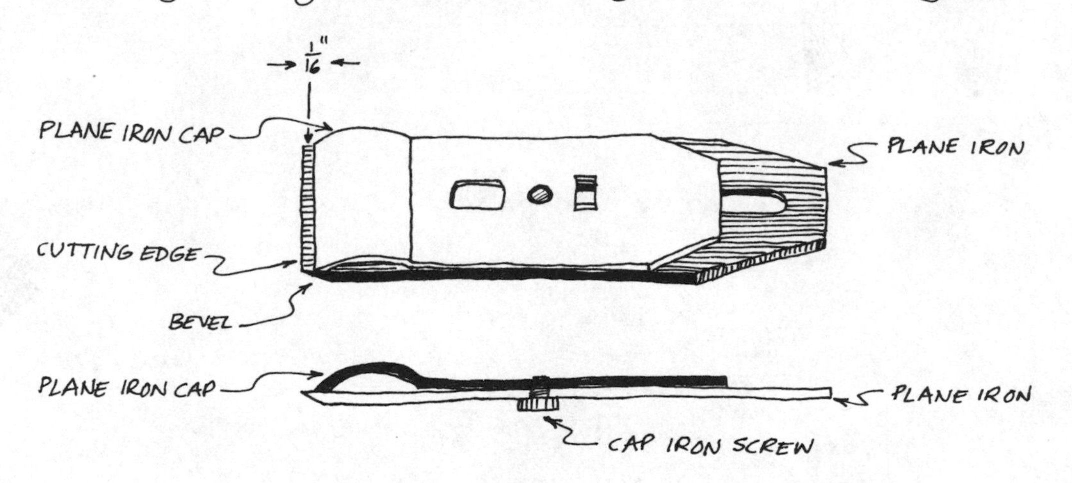

The plane iron cap is held tightly to the plane iron by the cap iron screw which is screwed in from underneath the plane iron, as shown below.

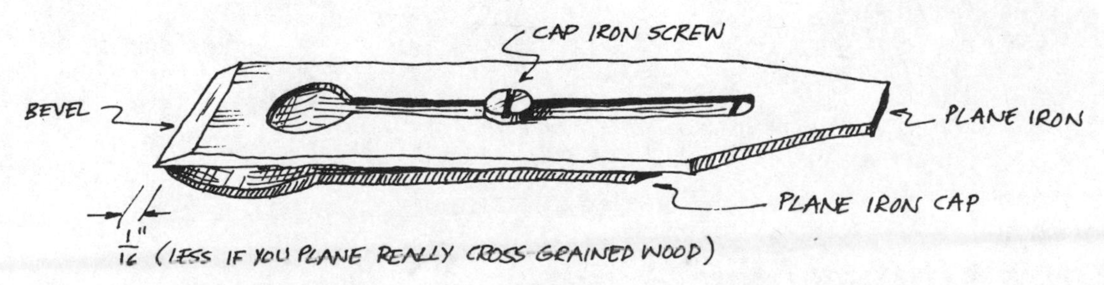

Adjustment of the double plane iron consists in making sure that the cap is set about $\frac{1}{16}$" from the edge of the blade. Sometimes the two irons must be disassembled to clean away shavings which sometimes clog things up. It is important to reassemble the double plane iron as shown below in order not to damage the cutting edge.

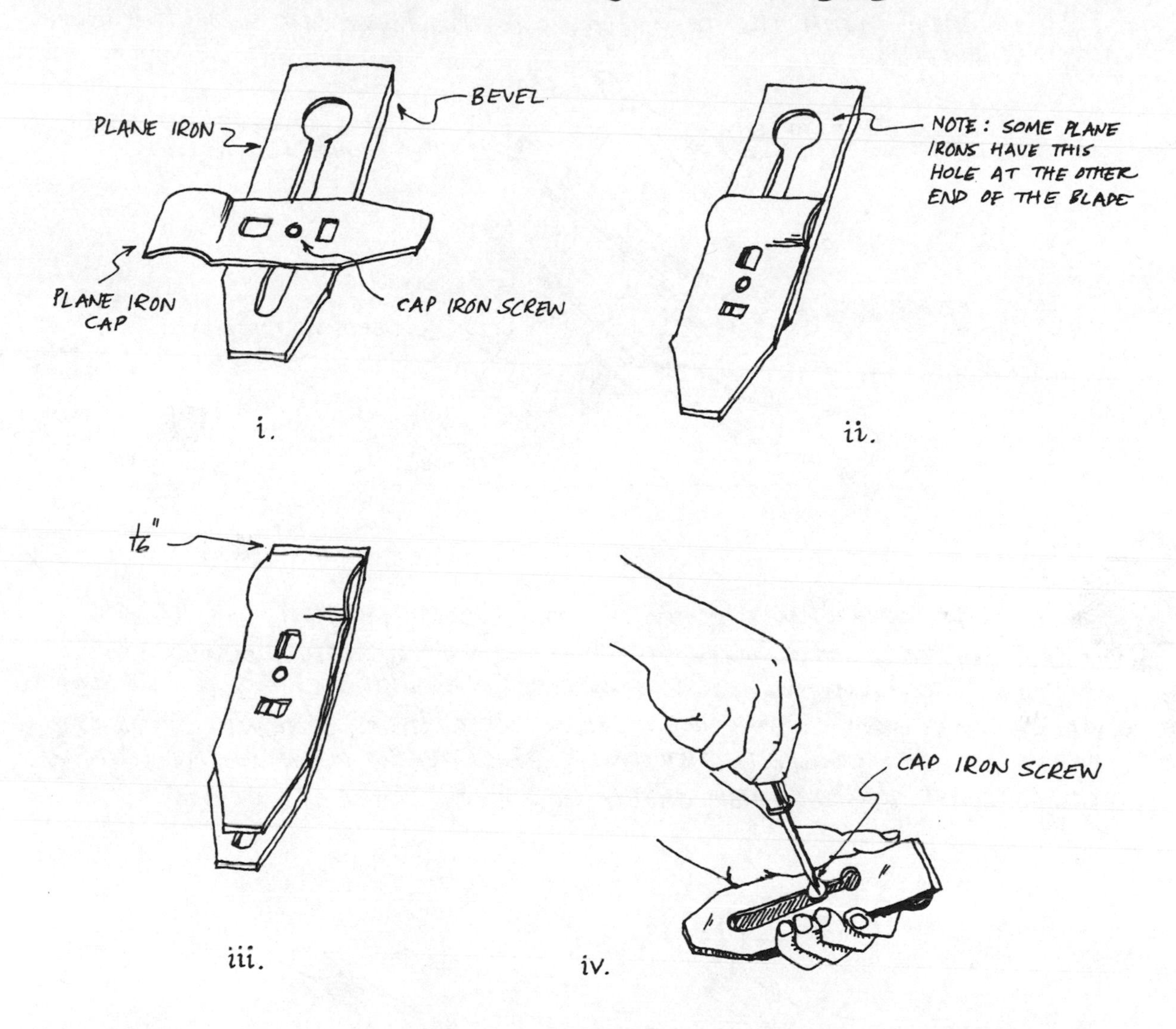

i. Put the plane iron cap on the flat side of the plane iron.
ii. Turn it straight on the plane iron.
iii. Slide the cap up the plane iron until just $\frac{1}{16}$" of the iron is left.
iv. Hold both parts tightly together, turn them over, and tighten the cap iron screw firmly.

2. THE LEVER CAP

Place the double plane iron into the body of the plane so that the lever cap screw and the end of the depth adjustment fit into the holes in the plane iron as shown. The bevel must be down, and the plane iron cap uppermost.

Then, with the cam up, slip the lever cap under the lever cap screw.

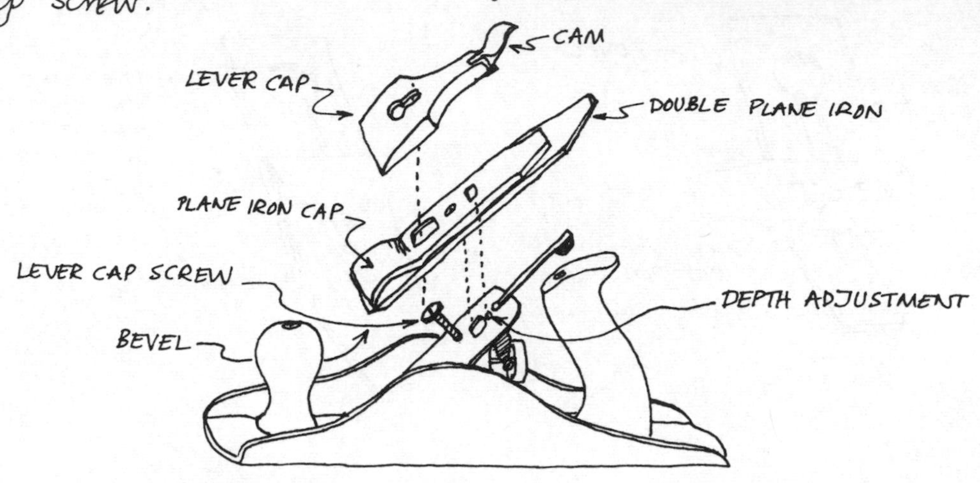

If everything is in the right position, you will be able to snap the cam on the lever cap down. If the double plane iron is not now securely held, you must take everything apart and tighten the lever cap screw — a quarter-turn at a time, clockwise, and try again. Conversely, if you must press really hard on the lever cap cam to get it flat, you should loosen the lever cap screw.

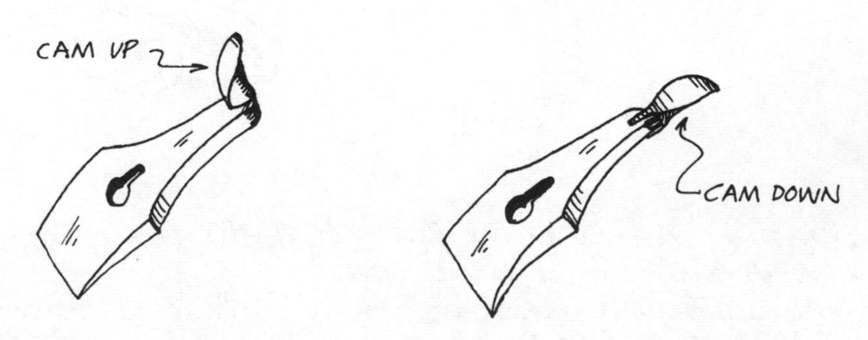

THE LEVER CAP

3. THE DEPTH OF THE CUT By turning the depth adjustment nut clockwise, the plane iron is drawn up. By turning it anti-clockwise, the blade is moved down and more will project through the opening in the bottom (or sole) of the **plane**.

In order to see what you are doing, turn the **plane** upside down and sight along the sole, holding the front (or toe) towards you.

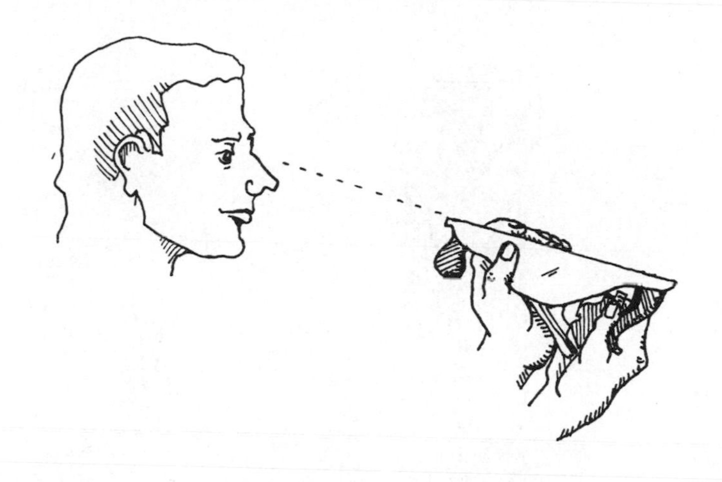

When you sight along the bottom of a correctly set **smooth plane**, or a **jointer plane**, the blade should project about the thickness of a hair. The shaving produced should be paper thin. **Fore planes** and **jack planes**, which are meant for heavier work, should have their blades projecting up to $\frac{1}{16}$" providing you can produce a smooth shaving without too much effort.

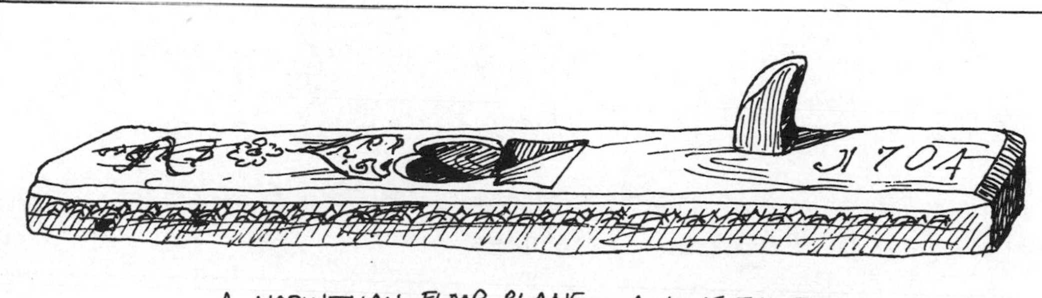

A NORWEGIAN FLOOR PLANE – A LONG JOINTER

4. THE LATERAL SETTING OF THE IRON For the plane to work well the cutting edge of the plane iron must project evenly through the bottom of the plane. This is achieved by adjusting the lateral adjusting lever.

Hold the plane as shown on the previous page and sight along the bottom again. As you move the lever from side to side this is what you should see:

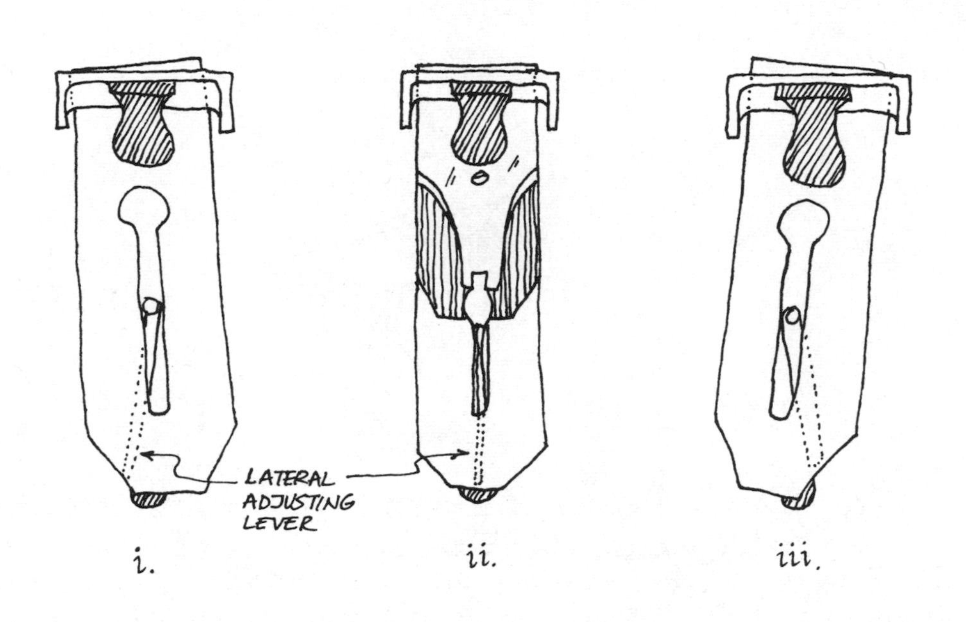

VIEWS i. AND iii. ARE WITH THE PLANE IRON CAP AND LEVER CAP REMOVED

i. When the lever is pushed to the left, the right-hand corner of the plane iron sticks up.

ii. When the lever is centered, the plane iron is more nearly equal.

iii. When the lateral adjusting lever is pushed to the right, the left-hand corner of the plane iron sticks up.

The lateral adjustment is greatly helped if you take care to hold the double plane iron as straight as possible when snapping the lever cap into place.

USING A BENCH PLANE

Make sure that what you are about to plane is secured. If you are at a bench for example, clamp the work in the **vise**, jam it up against the **bench stop**, or nail a V-board to the top of the bench.

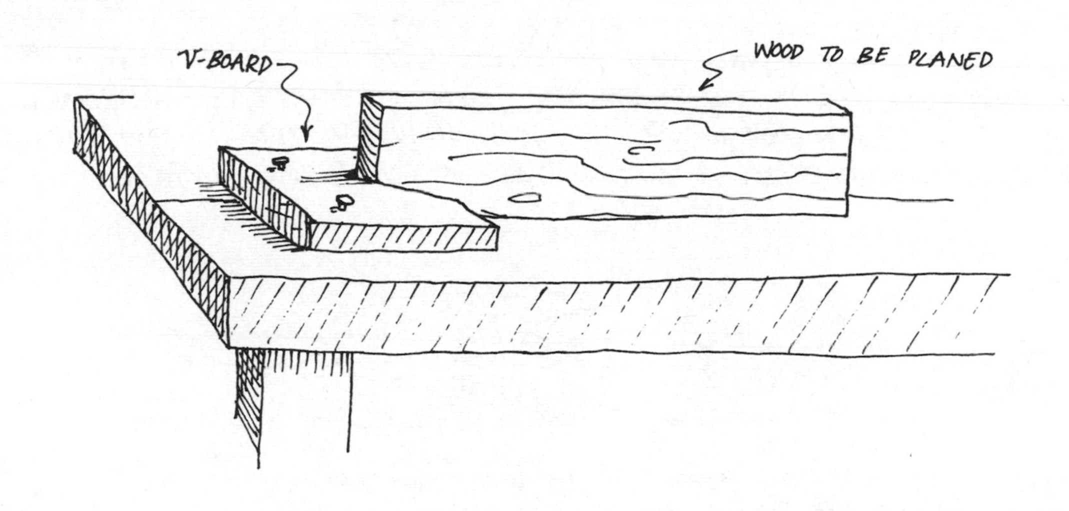

Try and plane the wood at hip height. Stand with your right side to the work and plane from right to left. Hold the handle with your right hand and place your left hand on the knob.

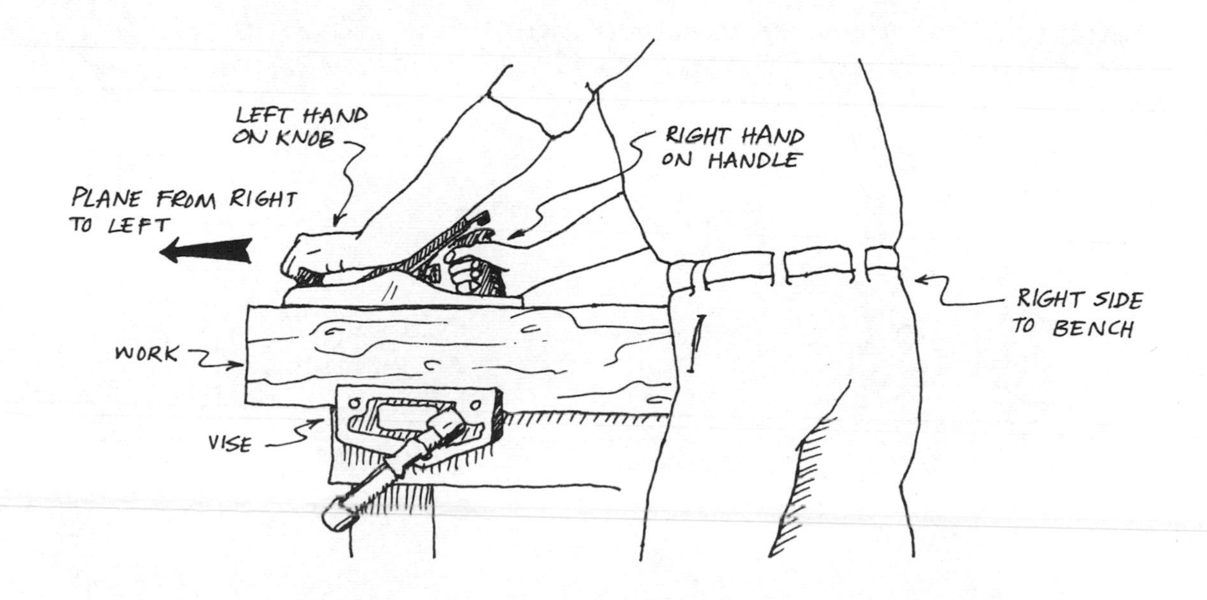

Always plane with the grain! If you don't, instead of smoothing the wood you will roughen it, and if you persist you may clog up the **plane**. If this happens, clean the **plane** and turn the wood around so that you can plane in the other direction. Sometimes you will have to plane the same piece of wood from both directions if the grain is changing.

It is not always easy to tell at first glance, although with experience you will be able to tell more easily, but sometimes looking at the <u>side</u> of the wood will help you follow the grain; you should plane in the direction that the grain comes to the surface:

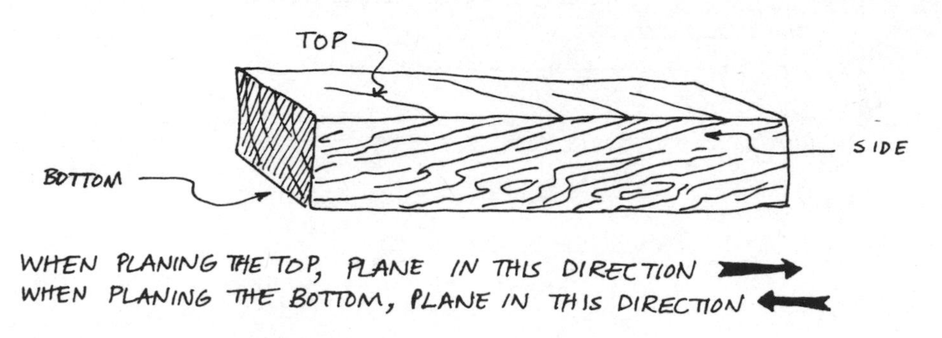

WHEN PLANING THE TOP, PLANE IN THIS DIRECTION →
WHEN PLANING THE BOTTOM, PLANE IN THIS DIRECTION ←

When you start to plane, press down on the knob; as you approach the end of the wood being planed, press down on the handle. This ensures that the wood is evenly planed throughout its length. On the return stroke, lift the back of the **plane** to protect the cutting edge.

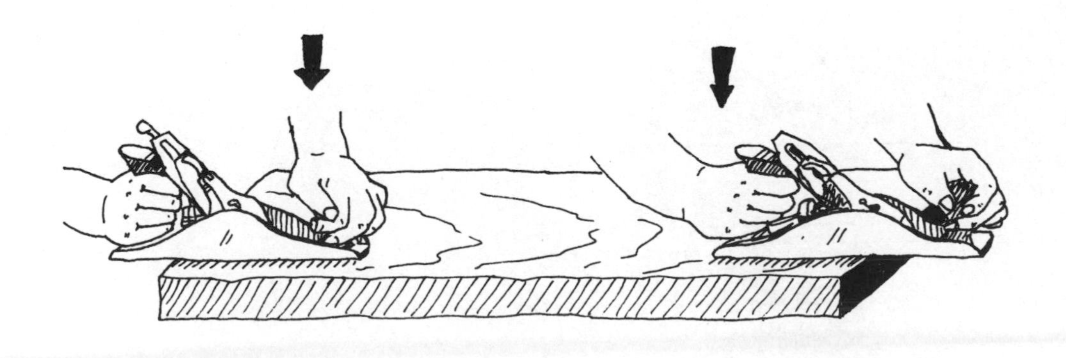

Do not try to plane too much at once - keep the shavings thin - it is easier work and you will not gouge the surface which you are trying to make smooth.

To plane the surface of a board, begin at one side and work slowly across the wood to the other side, using full-length strokes. To test how flat you have planed, place either edge of the sole of the **plane**, or the edge of a **square**, on the surface and see if any light shows underneath.

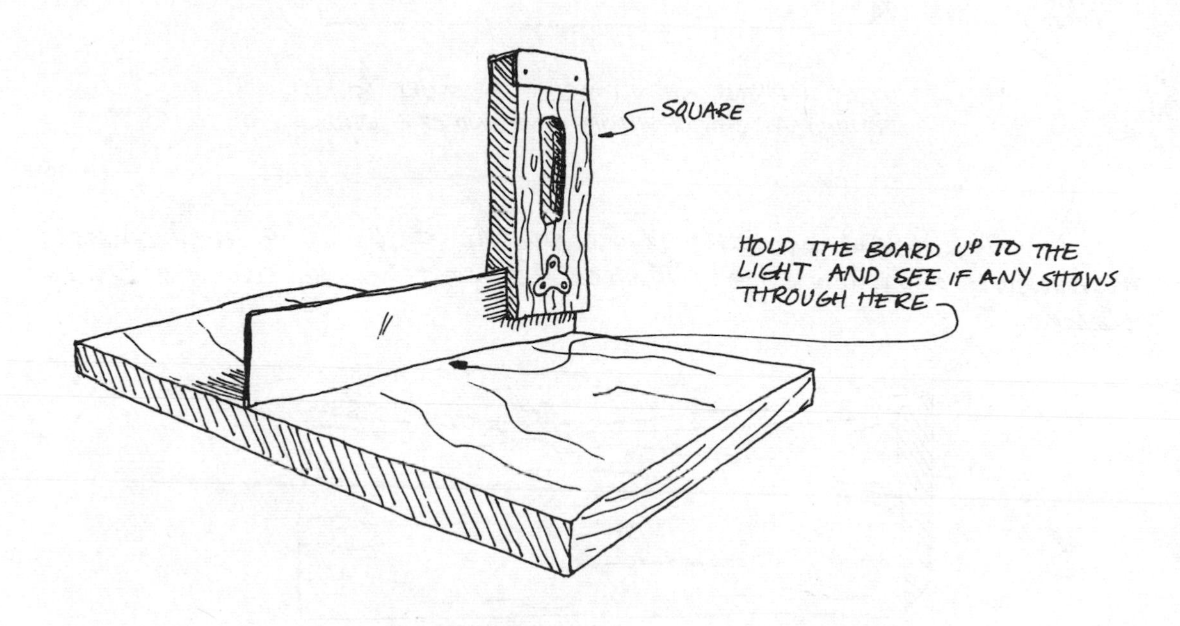

When not using it, lay the **plane** on its side to protect the cutting edge.

When planing an edge there are three common mistakes to avoid. The first one consists in planing the edge round because the correct pressure was not applied as explained on page 54.

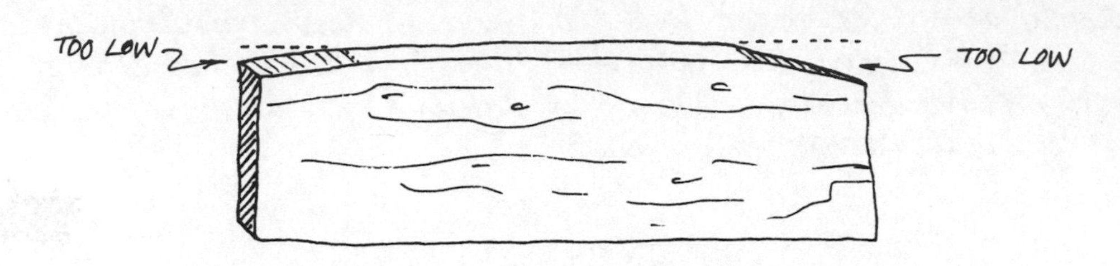

WHAT HAPPENS WHEN CORRECT PRESSURE IS NOT APPLIED AT THE BEGINNING AND END OF A STROKE

The second **mistake** is the opposite of the first and consists in making a hollow edge. This can be corrected by using a longer **plane**.

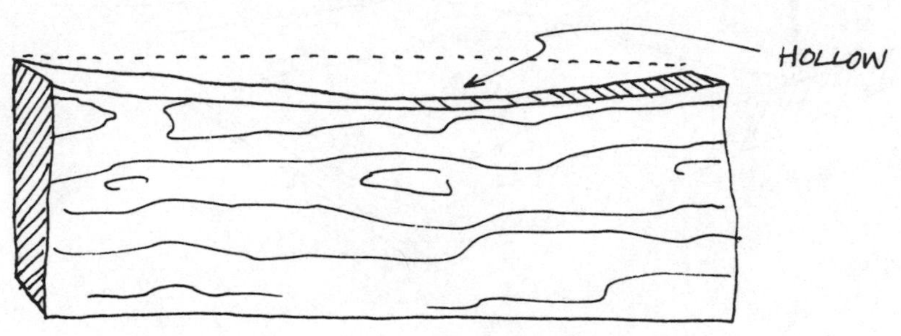

The third mistake is to plane the edge out of square.

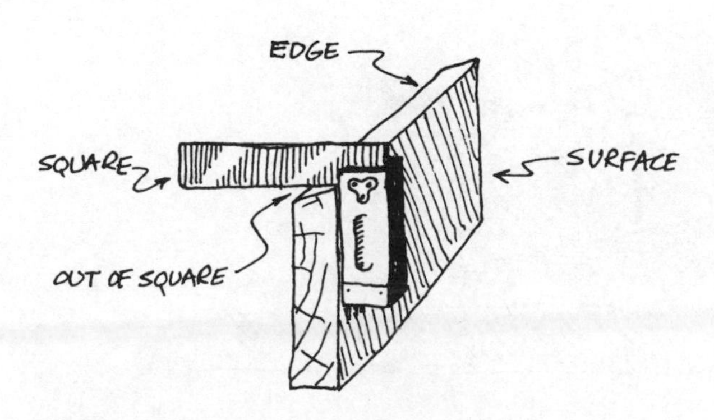

USING THE BLOCK PLANE

The block plane is meant for use with one hand. It only has a single plane iron and this should be placed with the bevel up – the opposite of bench planes.

Its main use is to plane end grain, for which purpose it has a very low angle plane iron.

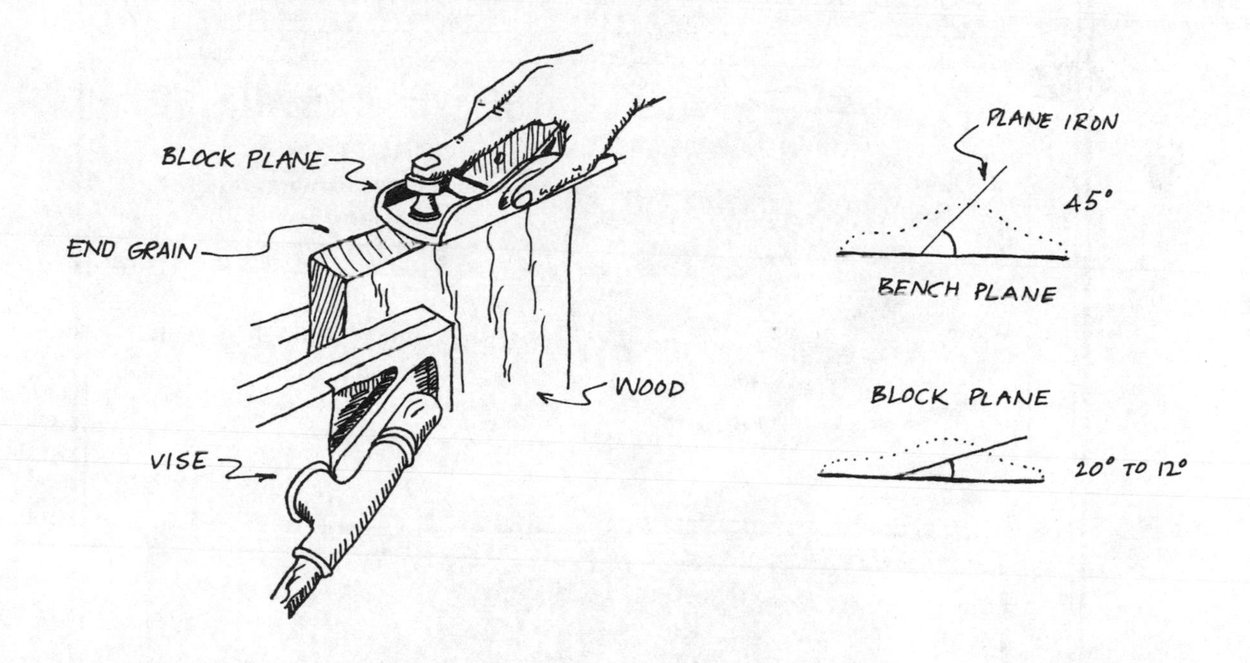

Care must be taken to stop before the end is reached when planing the end grain, or the far end will split.

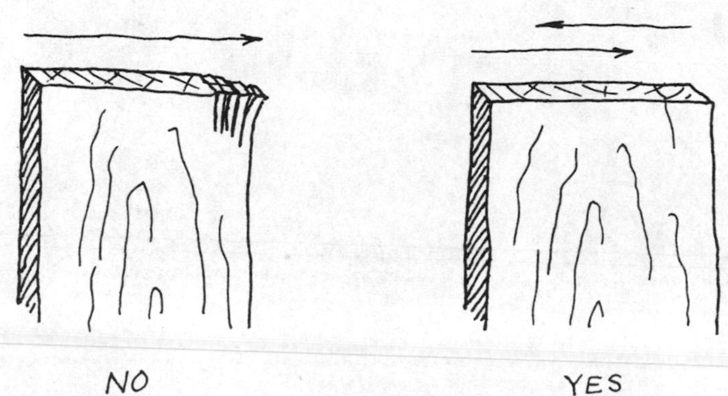

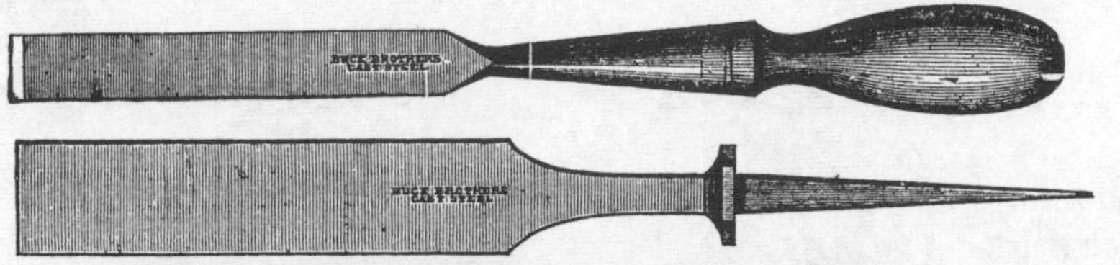

BUCK BROTHERS, Millbury, Mass.

The most complete assortment in the U. S. of

Shank, Socket Firmer and Socket Framing Chisels,

PLANE IRONS.

Gouges of all lengths and circles beveled inside and outside. Nail Sets, Scratch and Belt Awls, Chisel Handles. A full stock of Carving Tools. Also, small boxes of Tools of best quality.

Ten Eyck Axe Mfg. Co

COHOES, N. Y.

Warehouse, 103 Chambers St., N. Y.

Manufacturers of

AXES

Of all kinds.

Hatchets, Adzes, Grub Hoes, Mattocks and Picks.

Catalogues and Price Lists furnished upon application.

CHAPTER FOUR

Cutting —

NEVER CUT WITH A DULL TOOL, AND
NEVER CUT TOWARDS YOURSELF!

CUTTING TOOLS Tools with sharp edges used for separating wood are known as cutting tools. Among the many different tools which fall into this category are **adzes**, **axes** and **hatchets**, **knives** and **drawknives**, **planes**, which have been discussed in Chapter Three since they are so important, and, comprising the largest group of all, **chisels** - which very word comes originally from the Latin word cædere, meaning to cut.

ADZE An **adze** is like an **axe** with the blade at right angles to the handle. Although sawmills have made the **adze** obsolete in so far as hewing big house timbers, it is still much used by shipwrights.

SPUR
FOR HITTING NAILS
HANDLE
CUTTING EDGE

AXE The **axe** is surely one of the oldest tools known to mankind, and consequently has evolved into many specialised types, not all of them having to do with woodworking. Below is a picture of a two-handed wood **axe**.

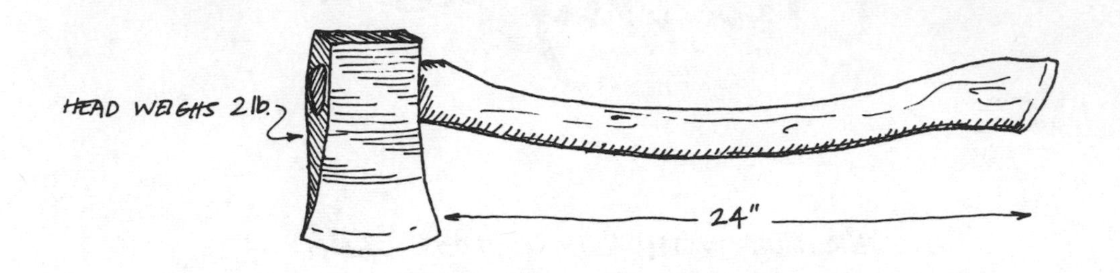

HATCHET A **hatchet** is literally a little **axe**; the word comes from the French word hachette, meaning little **axe**. It is much used by carpenters, and at one time there were many different types in common use, some of which are illustrated below.

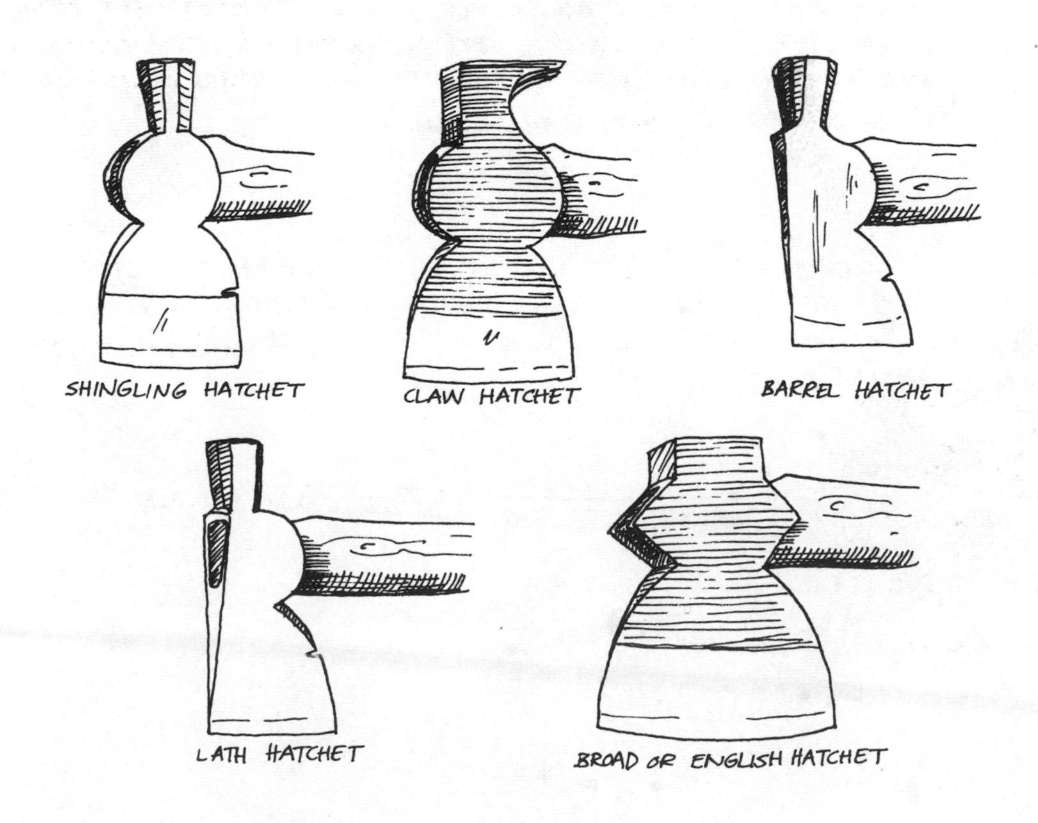

KNIFE The **knife** is probably the one tool with which everyone is familiar, and is made in every conceivable shape and form from **hunting knives** to **pocket knives**. Woodworkers use **pocket knives**, **sloyd knives**, and **utility knives**.

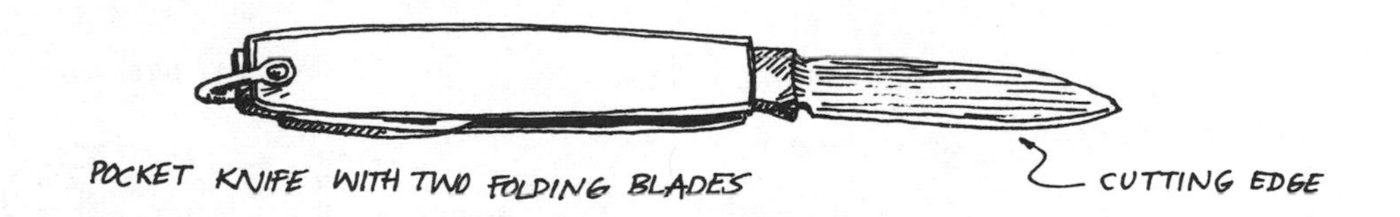

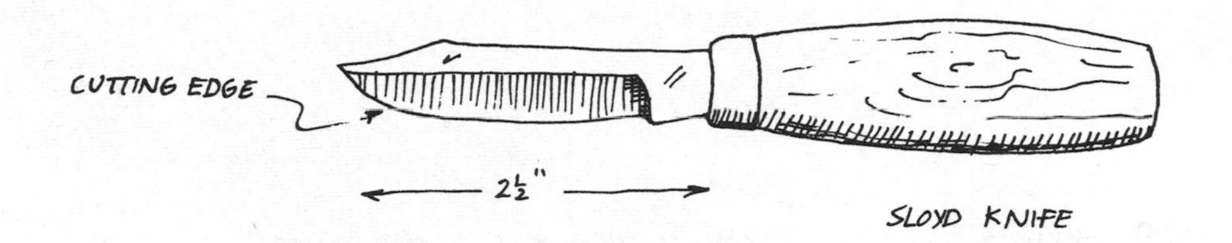

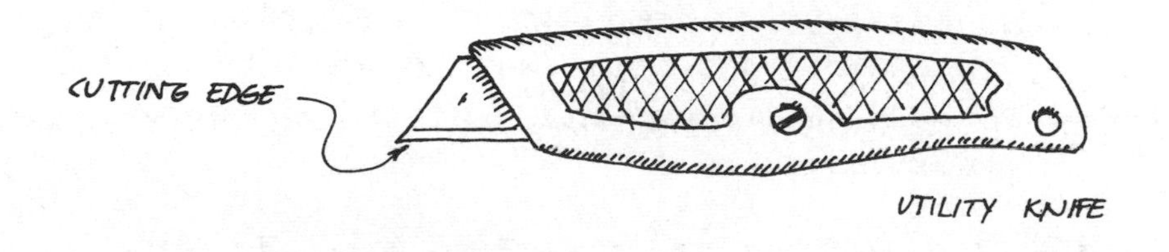

DRAWKNIFE A **drawknife** is a **knife** with a handle at both ends of the blade. It is called a **drawknife** because it is used by being drawn towards the operator. It is not so common now as when it was used by carriage makers, coach makers, wagon wrights, and other trades which have become obsolete, mechanized, or which have turned to materials other than wood.

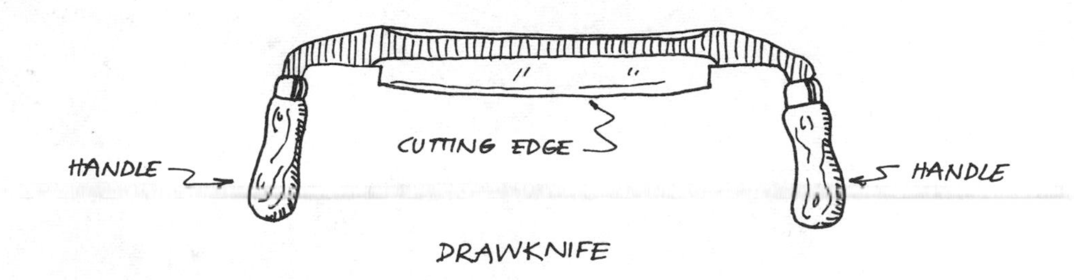

CHISEL A chisel is basically a metal blade held with a handle of wood or plastic, with the cutting edge at the end of the blade. There are many varieties of chisel, and most varieties are many times made in sets of differing sizes, some being as small as $\frac{1}{8}$th of an inch in width, and others being as wide as 2" or 3".

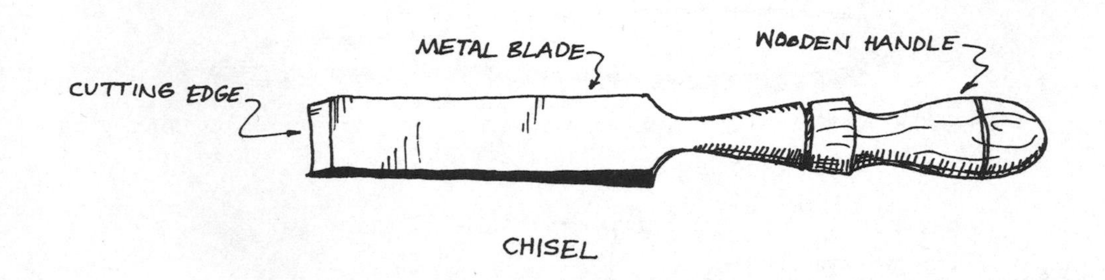

CHISEL

HOW CUTTING WORKS When a wood-cutting tool is very sharp, its edge divides the wood in its path, without either tearing the fibers apart or crushing them. If the edge is dull, it may make its way through the wood, partly by cutting and partly by tearing or crushing, but it will never make a bright clean cut when dull.

Also, the blade of a cutting tool is wedge-shaped, growing thicker from the edge to the back. As it penetrates, it forces the divided parts of the cut wood farther and farther apart.

If the chip or shaving removed is very thin, it bends easily out of the way of the tool, and does not hinder its progress.

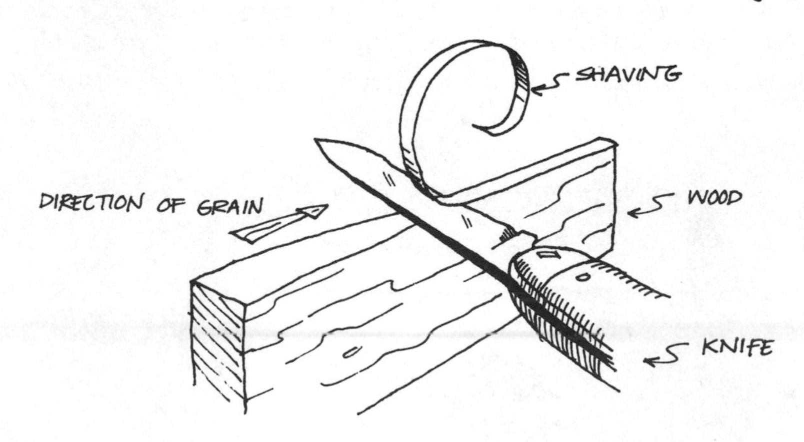

The same thing happens when a very thin shaving is taken crosswise of the grain as when the shaving is taken lengthwise – as was illustrated on the previous page.

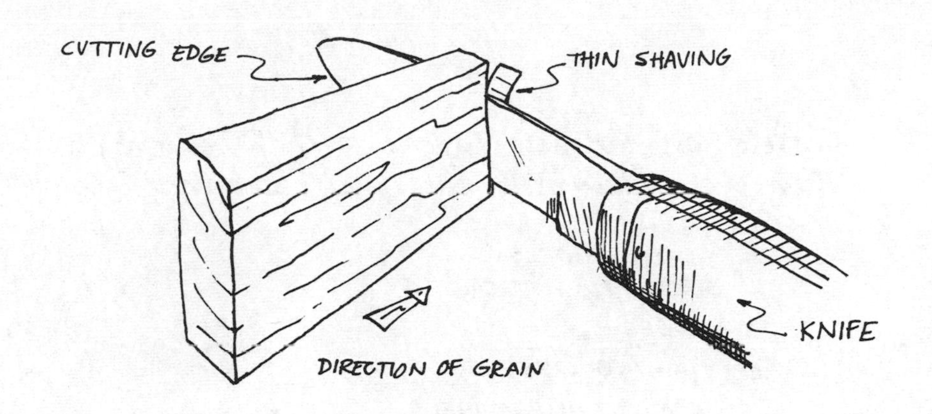

In both cases the shaving is so thin that it moves easily out of the way and the wedge has practically no work to do.

If, however, the chip or shaving is so thick that it does not bend out of the way, then the wedge of the tool compresses the sides of the wood and makes room for the cutting edge as it cuts deeper and deeper. When the wood is much compressed, great force is required to make a cut.

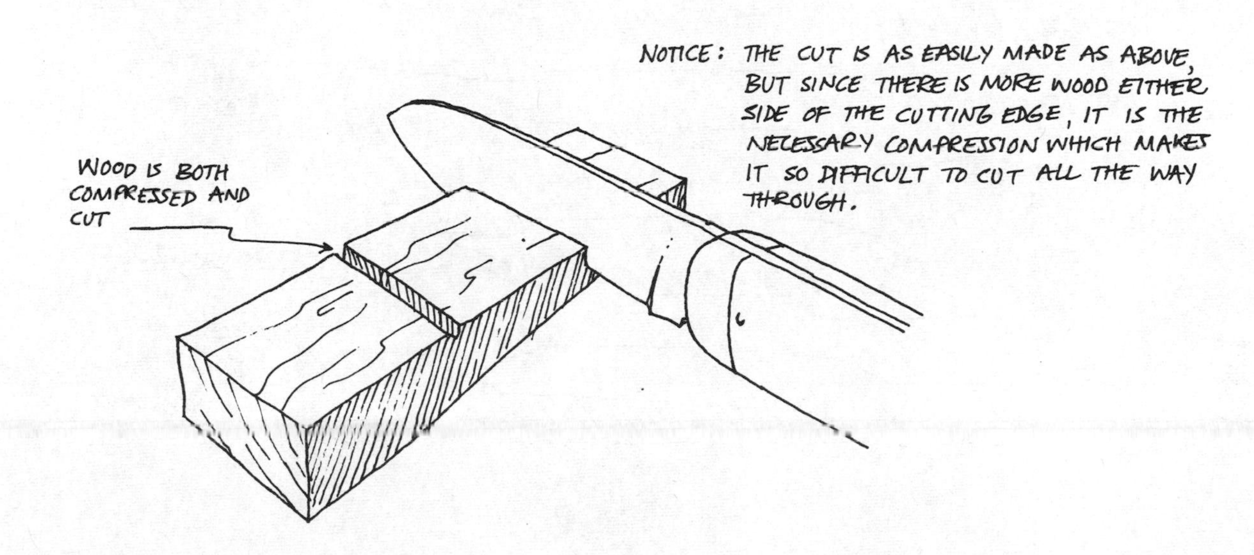

If even the very sharpest cutting edge is examined through a powerful microscope, it will be found to be notched, like a tiny **saw**. It is for this reason that a cutting edge will cut more easily, and make a cleaner cut, if the edge is slid in the direction of its length while cutting.

CHISELS

Chisels are the most important of the cutting tools. There are many different kinds, but they may be grouped into five major classes:

1. BENCH CHISELS
2. SPECIAL-PURPOSE CHISELS
3. GOUGES
4. TURNING CHISELS
5. WOODCARVING CHISELS

1. BENCH CHISELS

These are the common tools used for most woodworking operations. Yet even these may be further divided into 3 main groups:

i. **paring chisels**
ii. **firmer chisels**
iii. **framing chisels**

The difference is chiefly one of heaviness of work, the **paring chisel** being the lightest and thinnest, the **framing chisel** being used for the heaviest work.

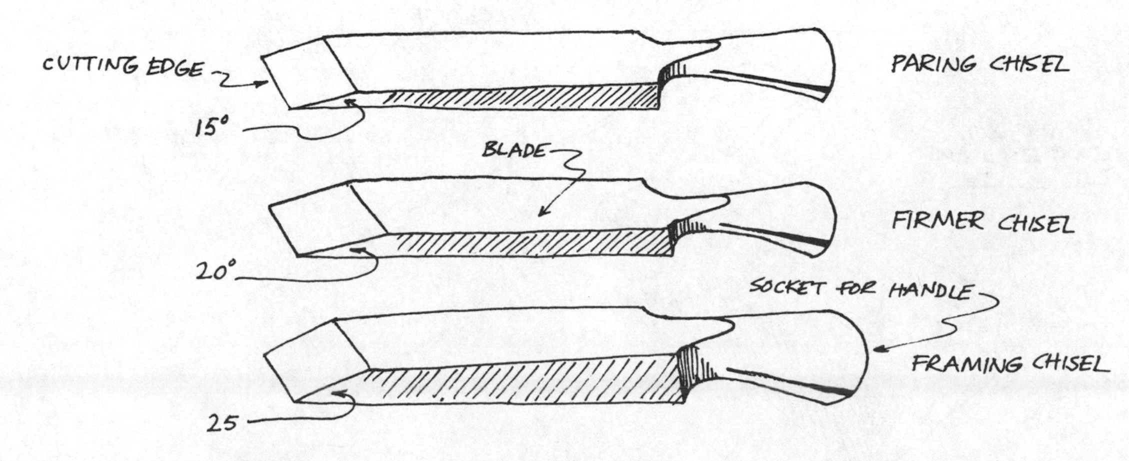

Most **chisels** may also be classed as either "tang" or "socket," according to how the handle is attached.

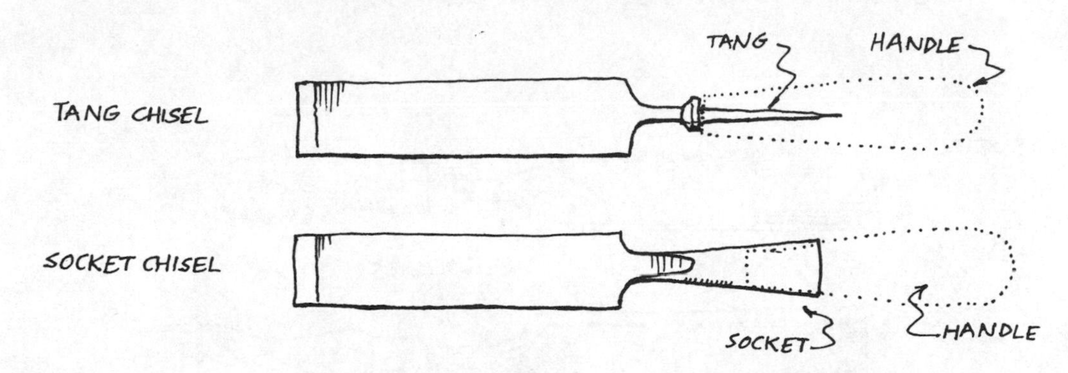

Generally, light **chisels**, such as **paring chisels** and small **woodcarving chisels**, are of the tang variety, while **chisels** intended for heavier work, where they will receive **mallet** blows on the handle, such as **firmer chisels**, are usually of the socket type.

2. SPECIAL-PURPOSE CHISELS

Chisels belong to this group, which don't belong to any of the other four groups.

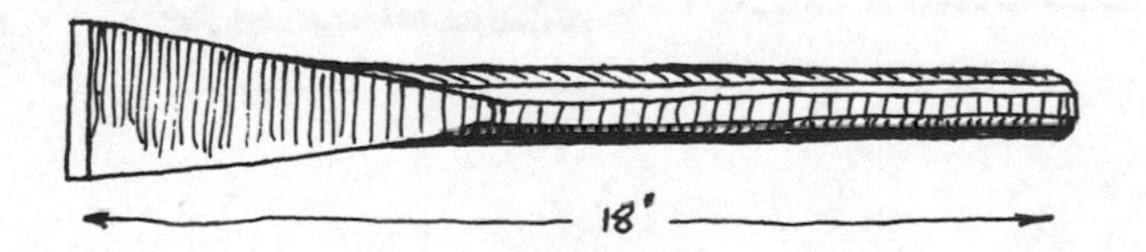

A FLOOR CHISEL, USED TO CUT FLOORING

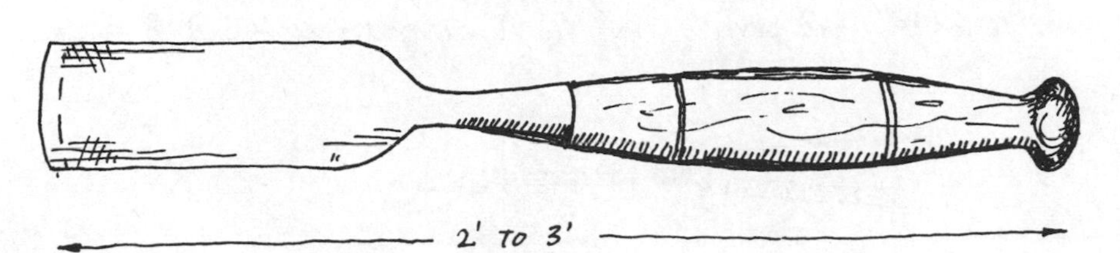

A SLICK, A TWO-HANDED, GIANT PARING CHISEL, USED IN BOAT BUILDING

3. GOUGES A *gouge* is a *chisel* with a hollow-shaped blade for cutting curves or round holes. The cutting edge, or bevel, may be on the outside or the inside of the curve.

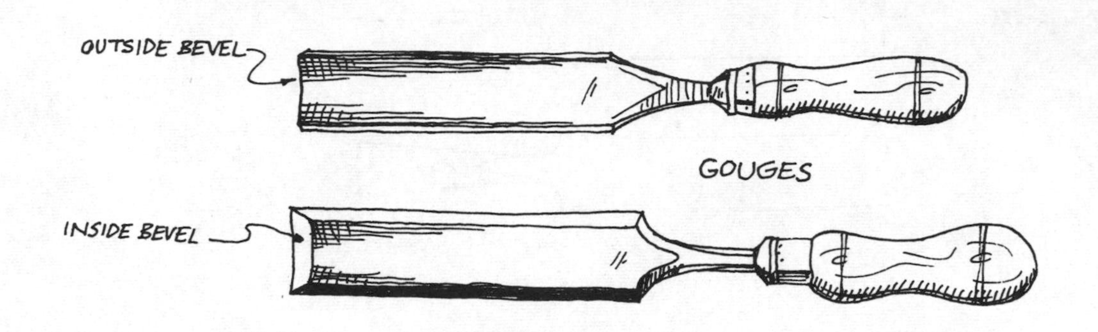

4. TURNING CHISELS *Turning chisels* are those *chisels* designed and used for work with a wood-turning lathe. They are generally of the tang variety and are characterized by having longer blades and longer handles than most other *chisels*.

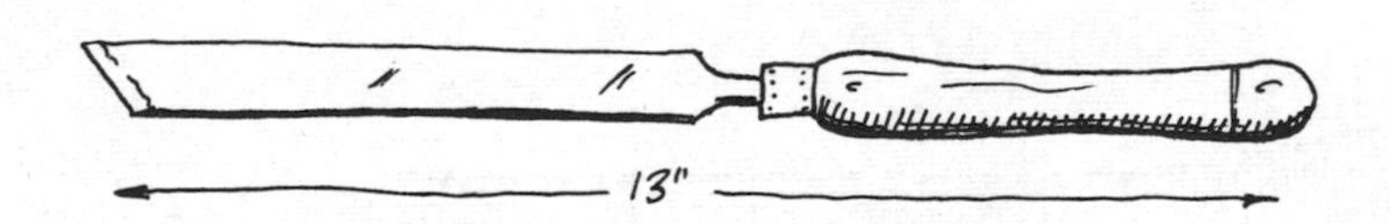

5. WOODCARVING CHISELS *Woodcarving chisels* are made in a great variety of shapes, curves, and angles. They usually have tangs, and very often have octagon-shaped handles.

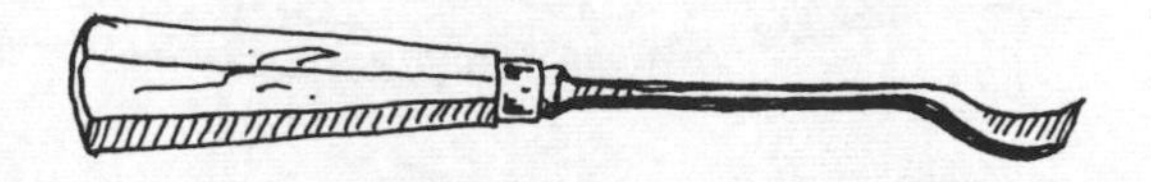

SHORT-BEND WOODCARVING CHISEL

HOW TO USE CHISELS

For good work, done with the least effort, observe these few simple rules:

1. Keep the tool bright and sharp at all times.
2. Protect the cutting edge. Do not allow the **chisel** to touch other tools. Always lay the tool down with the bevel side down.

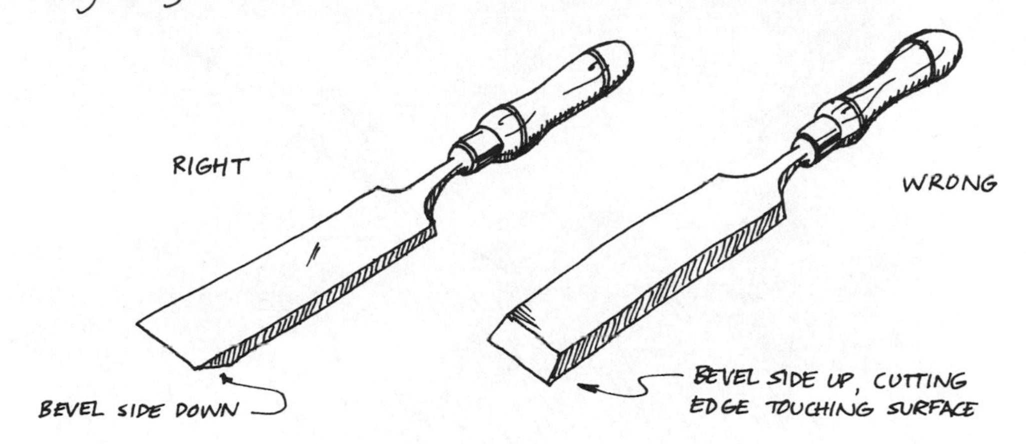

3. Don't use **chisels** as **screwdrivers**, or to pry things open.
4. Hold the work in a **vise** whenever possible in order to have both hands free to control the **chisel**.
5. Always cut <u>away</u> from you and keep your hands <u>behind</u> the blade.
6. Use the left hand to guide the **chisel** and the right hand to push the handle forward.

When chiseling with the grain:

1. Hold the bevel down for big cuts; hold the bevel up for thin "paring" cuts.

BEVEL

GRAIN

FOR DEEP CUTS : BEVEL UP

HOLD THE CHISEL AT AN ANGLE TO OBTAIN THE SLIDING EFFECT MENTIONED ON PAGE 64

GRAIN

BEVEL

FOR THIN CUTS : HOLD CHISEL FLAT ON WOOD WITH BEVEL UP

2. Always try to chisel with the grain, as in planing, to avoid splintering the wood.

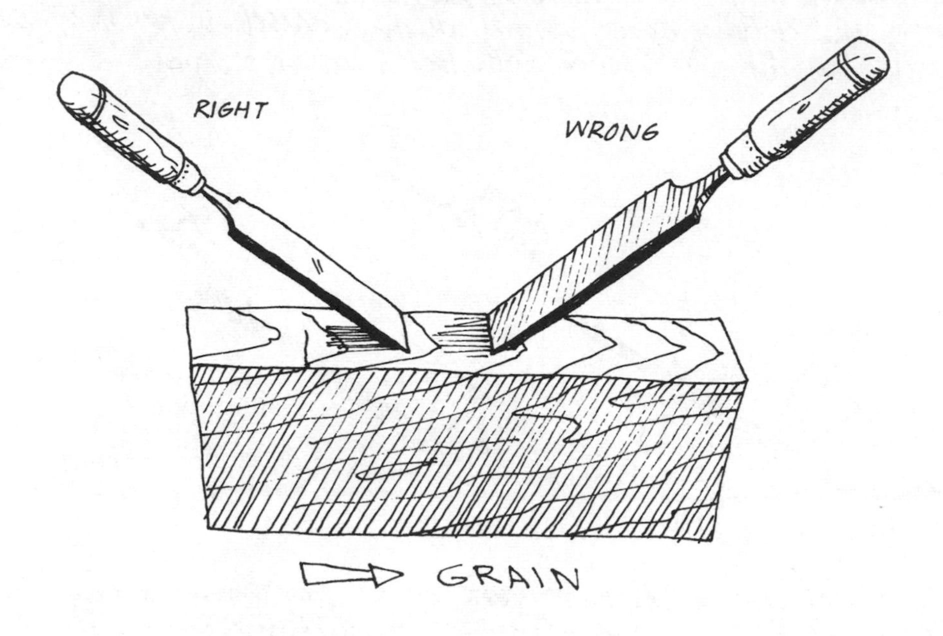

When chiseling across the grain:

1. Hold the blade between the thumb and first two fingers of your left hand to guide and act as a brake while you push with your right hand.

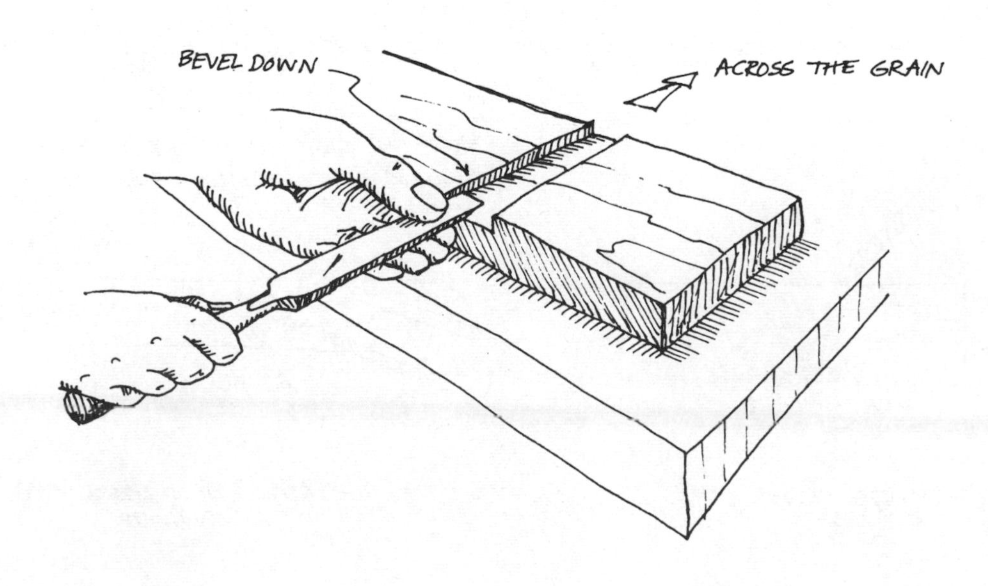

2. For heavier chiseling use a **mallet**, never a **hammer**, unless your **chisel** has a plastic handle.

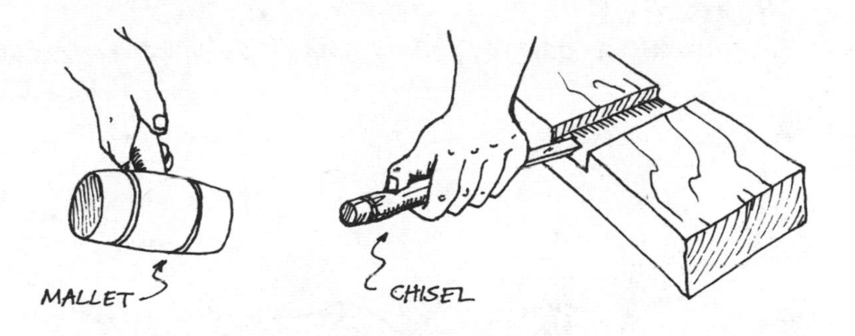

3. Do not cut all the way across a board from one side to the other, but work towards the middle from both sides; this avoids splintering.

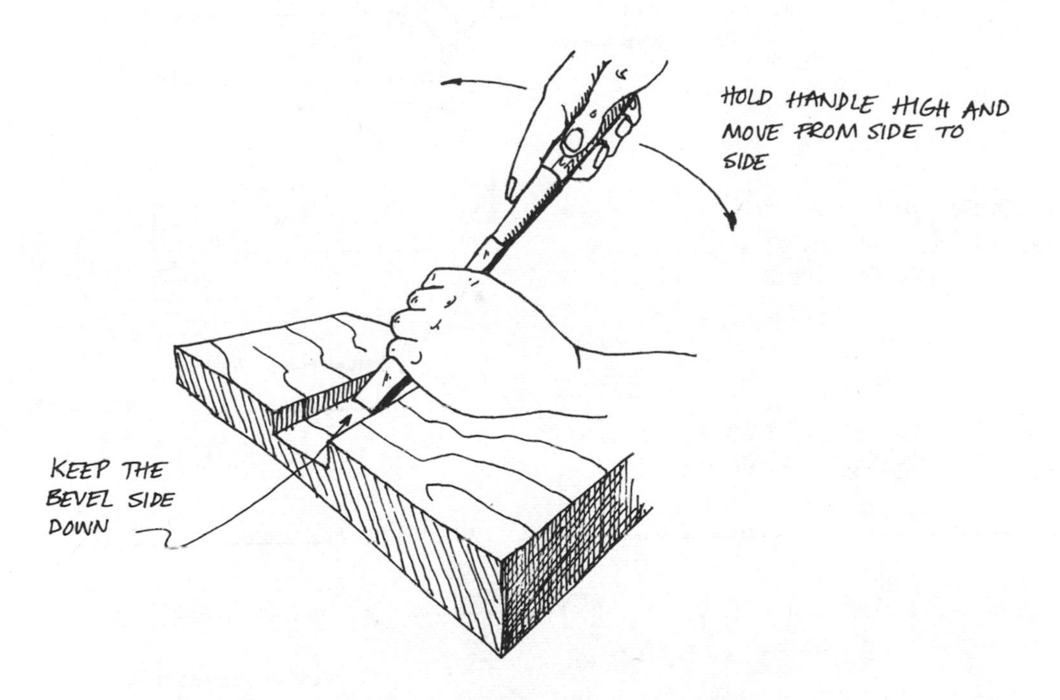

When chiseling across wide boards, where the **chisel** will not reach all the way across, you must raise the handle of the **chisel** - but keep the bevel down - and move the handle slightly from side to side. This gives the blade an oblique cutting motion, which gives a cleaner cut.

When chiseling end grain:

1. Don't try to chisel off more than $\frac{1}{8}$th". Use a **saw** to remove the bulk, then chisel.
2. Start to cut at a corner, and move obliquely across the board, as at **1**.

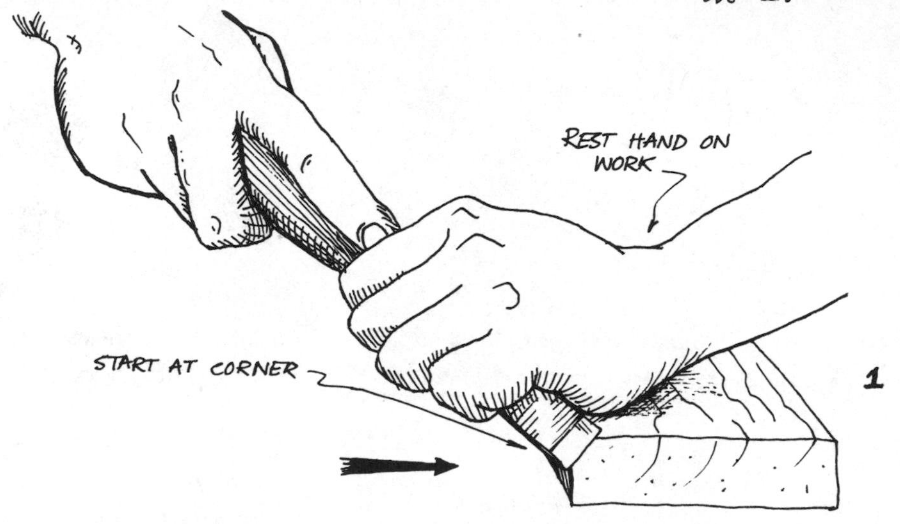

As the stroke proceeds, lift the handle, until, by the end of the stroke, the **chisel** is nearly vertical, as at **2**.

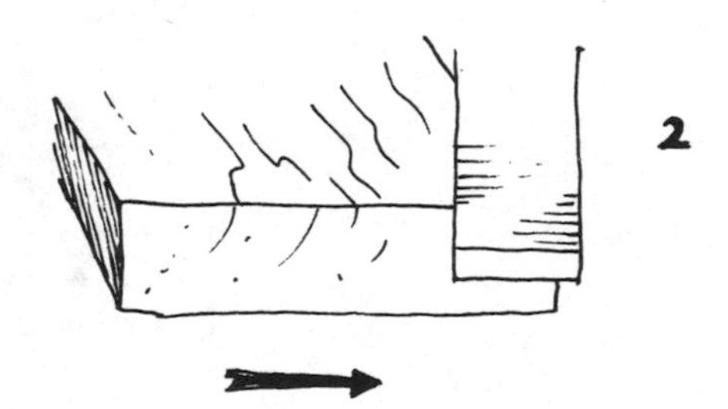

3. Keep the back of the blade flat against the wood just chiseled.

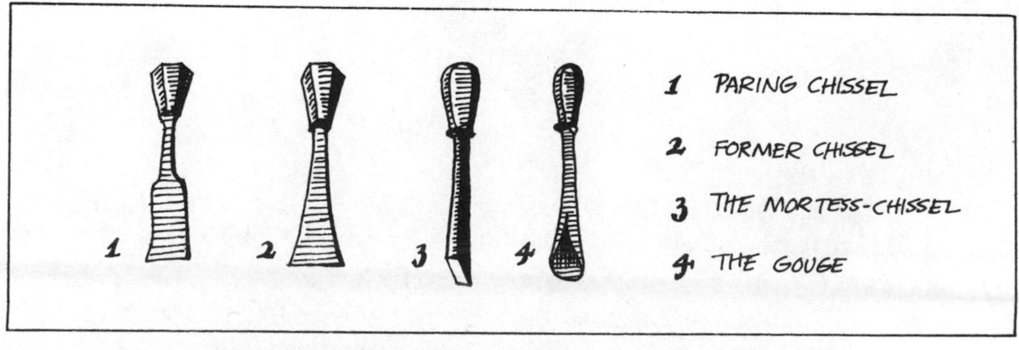

CHISSELS OF SEVERAL SORTS, AFTER MOXON 1703

TO CUT AN INSIDE CURVE

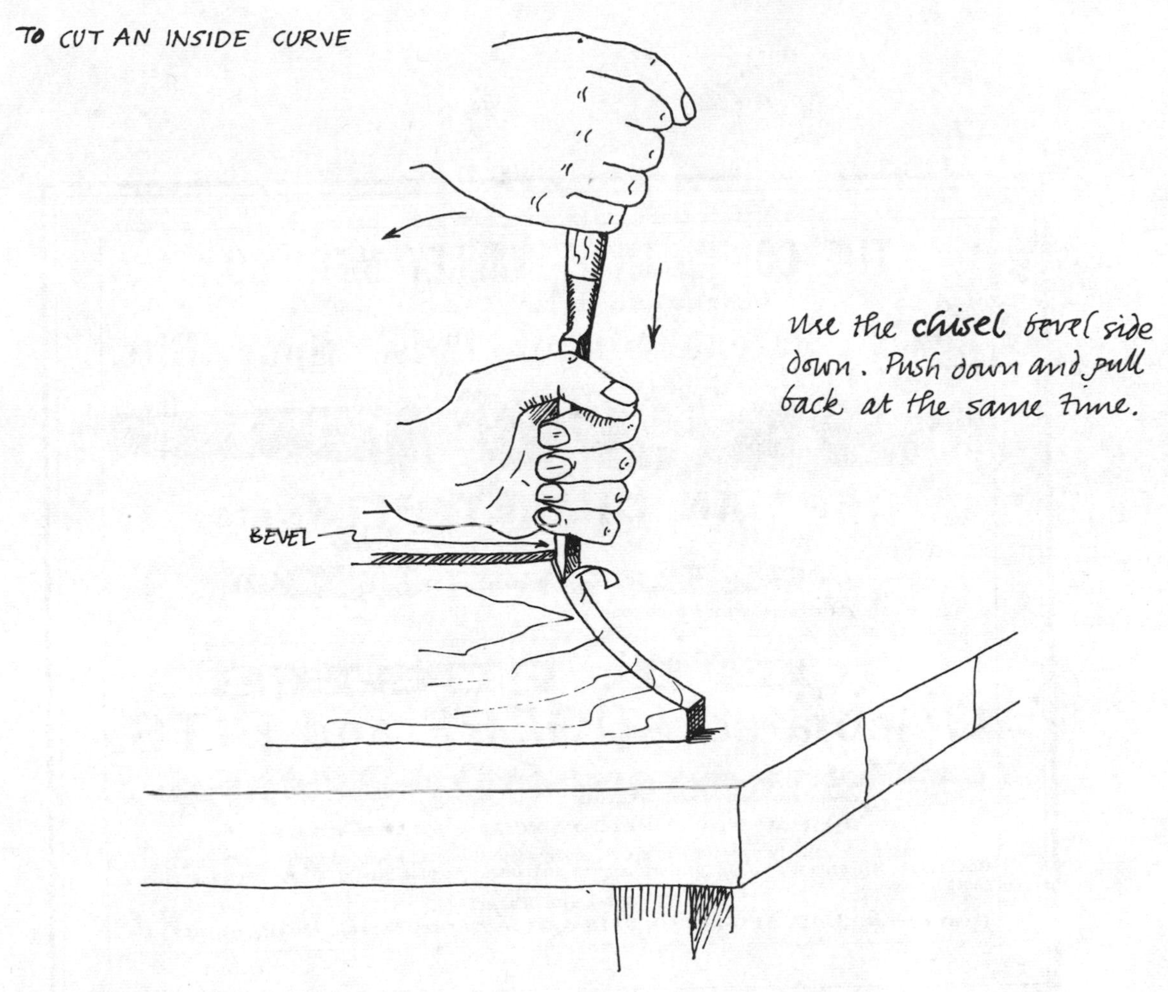

Use the **chisel** bevel side down. Push down and pull back at the same time.

TO CUT AN OUTSIDE CURVE

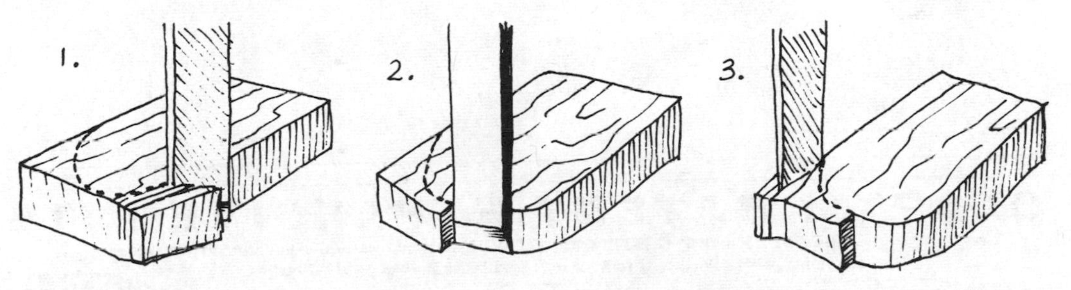

Cut to the line as shown. The first cut should be made a little outside the line, and near one side. The chips will then easily split out of the way. After reaching the middle of the curve (2.), begin again at the other end. Finally, finish with light chips, cutting as close as possible to the line.

THE CONNECTICUT VALLEY MFG. CO.,

CENTERBROOK, CONN., Manufacturers of

Lewis' Patent Single Twist Spur Bits,

GERMAN GIMLET BITS, etc.

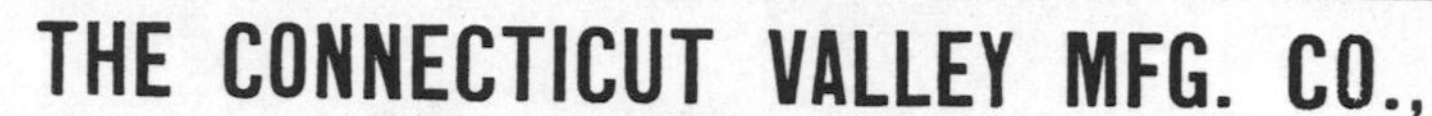

Send for our illustrated price list and discounts.

JOB T. PUGH'S

Celebrated AUGERS and BITS.

WARRANTED SUPERIOR TO ANY OTHER MAKE.

They are made entirely by hand, and are especially adapted to hard wood. Supplied to the trade only. Gas Fitters', Millwrights', and Carpenters' Augers and Bits. Machine Bits of all descriptions made at short notice.

Office and Works,

Rear of Nos. 3112, 3114, 3116, 3118 & 3120 Market Street, Philadelphia, Pa.

CLARK'S PATENT EXPANSIVE BITS

Made of JESSOP'S BEST CAST STEEL, and warranted superior to any other

Two sizes: Large Size Boring, ⅞ to 3 inches ; Small Size Boring, ½ to 1½ inches.

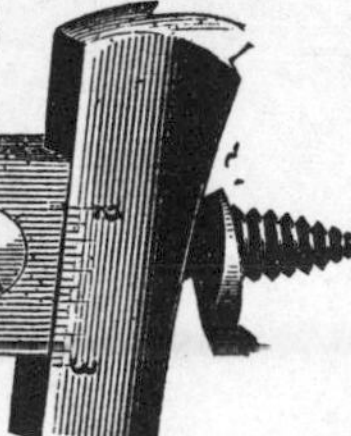

Manufactured by

WILLIAM A. CLARK, - - - Westville, Conn.

CHAPTER FIVE

Boring ~

PAY CONSTANT ATTENTION TO HOW STRAIGHT YOU BORE!

Boring is one of the more difficult woodworking operations (even though in essence it is very simple), because it is only with practice and good tools that satisfactory results may be obtained.

BORING TOOLS There are very few tools for boring, even though some of them have been known for thousands of years. Basically, they may all be classified into one of the four following groups:

1. AWLS
2. GIMLETS
3. DRILLS
4. AUGERS

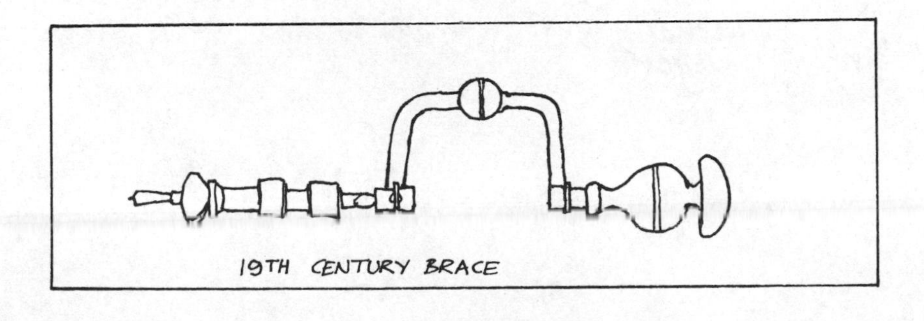

19TH CENTURY BRACE

1. AWLS The **awl** most commonly used by the woodworker is the **brad awl**. It is generally shaped like a small **screw-driver**, and, in fact, may be used as such for very small screws. Its chief purpose, however, is to make small starting holes for other **drills** or screws. Apart from being careful when you place it while making a small hole, the only rule connected with the **brad awl** is that it should be started with its blade across the grain, as shown.

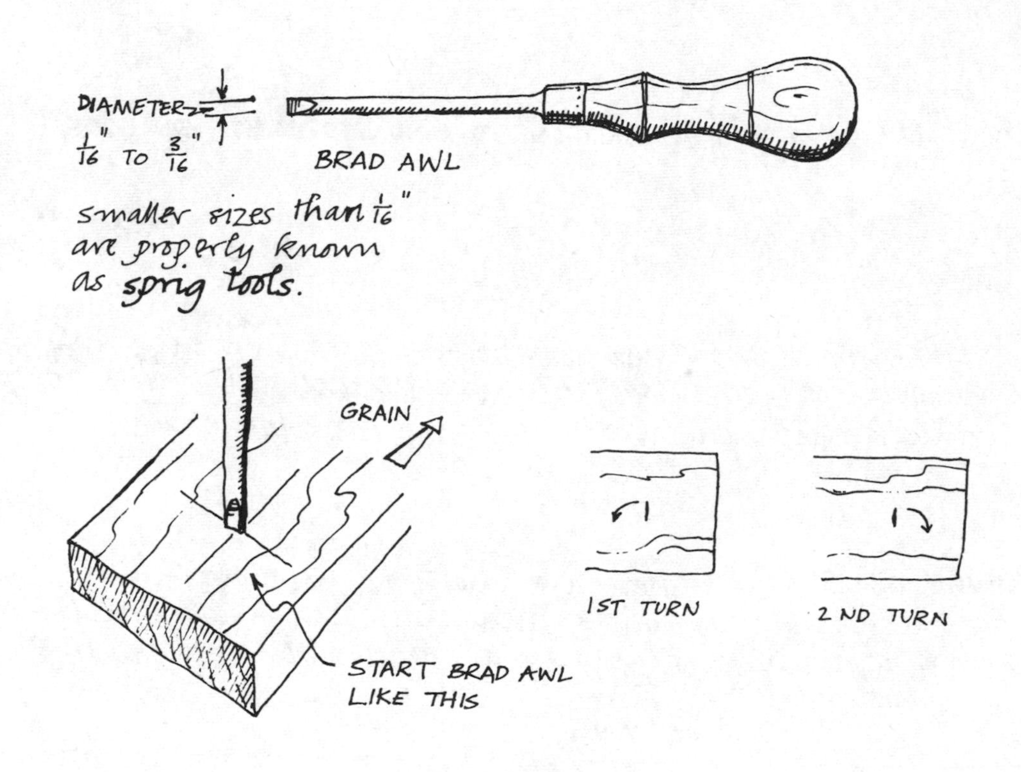

2. GIMLETS The word **gimlet**, sometimes spelled **gimblet**, means actually just a small **wimble** - which was the medieval word for **auger**.

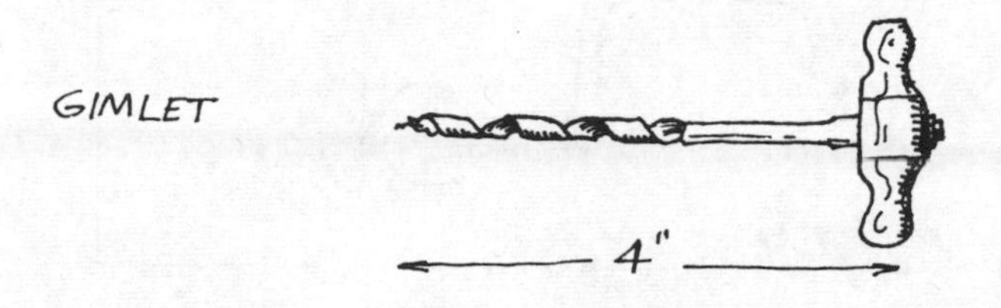

The **gimlet**, while used like a **brad awl** to make small starting-holes in wood for screws, is actually a small **auger** and may be used for drilling small holes by hand-pressure. In fact, large **gimlets** - from 1/4" to 1/2" in diameter - are known as **auger-gimlets**.

To use the **gimlet** simply hold it in the right hand, and, while pressing on the handle with the palm of your hand, drive it into the wood with a series of half turns until it is as deep as needed.

3. DRILLS There are three **drills** commonly used by the woodworker. They are the **hand drill**, the **breast drill**, and the **push drill**. These tools work by turning variously sized **drill bits** in the wood. It should be pointed out that the metalworker uses slightly different nomenclature for the same tools. He refers to the **drill bit** as the **drill**, whereas to the woodworker, the **drill** is the tool which drives the **bit**.

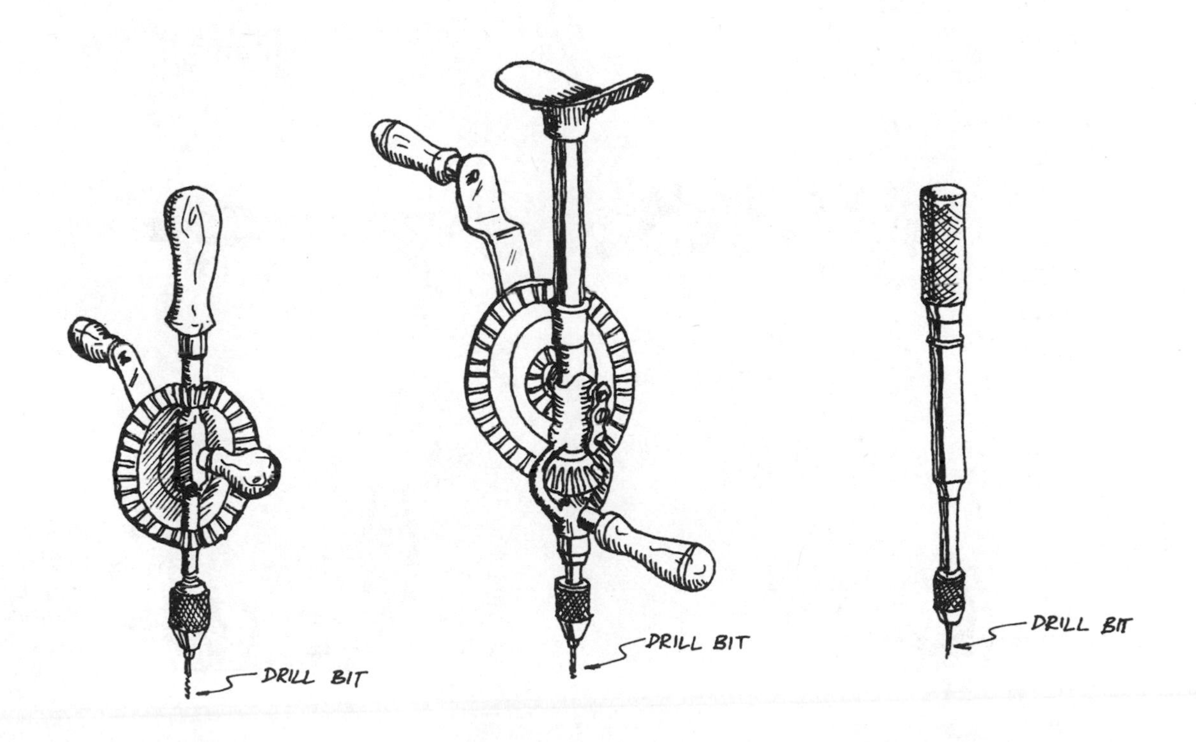

HAND DRILL BREAST DRILL PUSH DRILL

4. AUGERS An **auger** is properly a complete tool, comprised of a handle and a bit, like the **carpenter's nut auger** shown below. However, since these tools are now rarely used, the word **auger** is now most commonly used to refer to what is more correctly an **auger bit**, which, like the **drill bit** mentioned on the previous page, requires another tool to drive it; in the case of the **auger bit**, the tool needed is the **brace**.

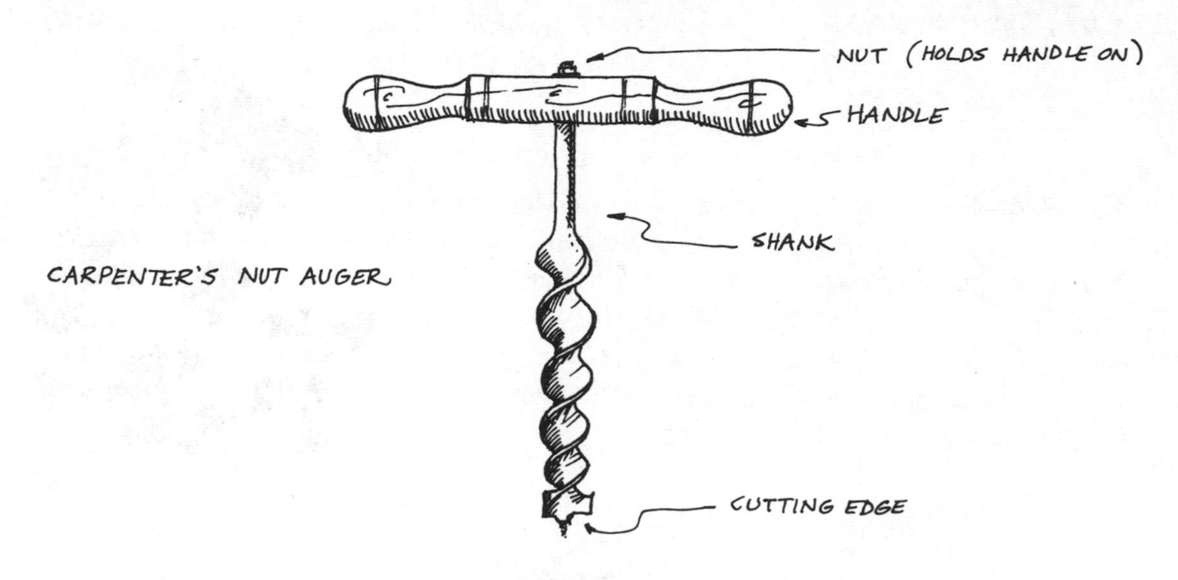

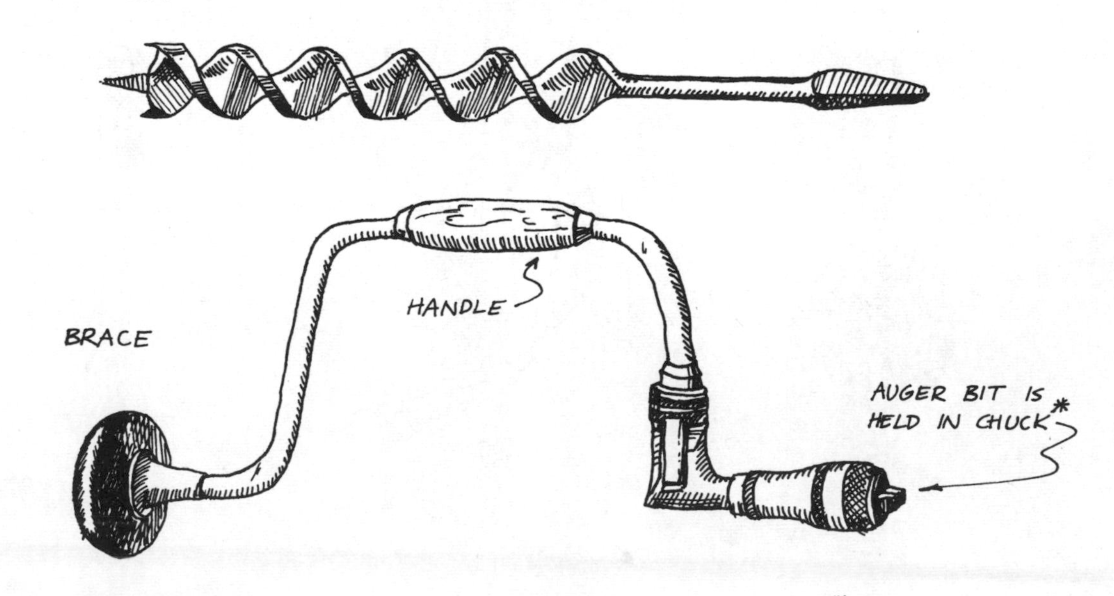

* SEE PAGE 81

HOW TO USE DRILLS

Hand drills are used for boring small holes in wood. They are small, light, and generally more convenient than **braces**. Since **drill bits** do not have feed screws like **auger** bits they are not pulled into the wood and there is, consequently, less danger of the wood splitting.

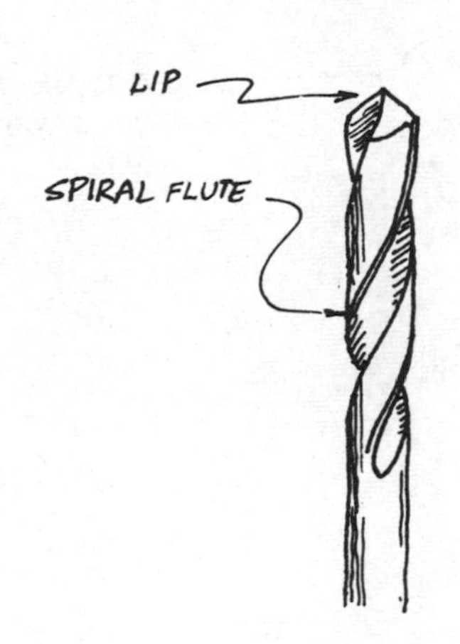

THE END OF A DRILL BIT

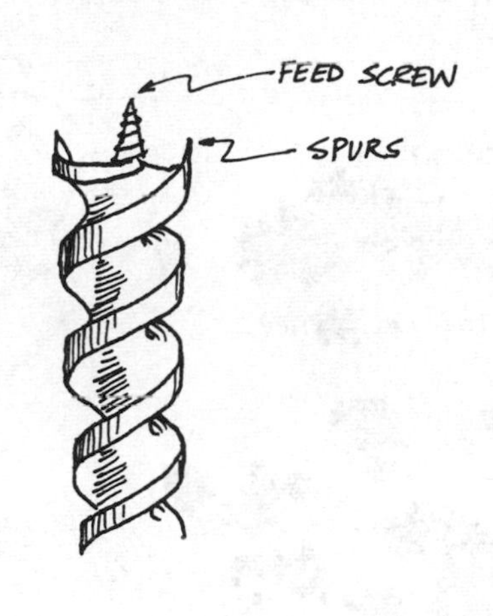

THE END OF AN AUGER BIT

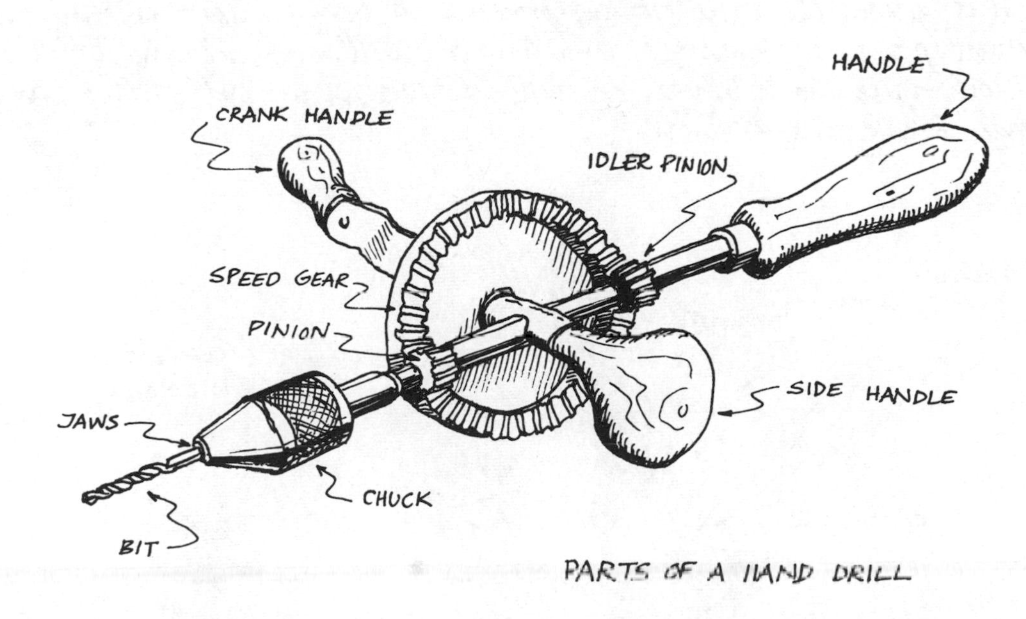

PARTS OF A HAND DRILL

To open the chuck, into which you place the **drill bit**, turn the crank handle backwards while holding the chuck firm. Do not open the jaws in the chuck more than is necessary to insert the **bit** - it will be easier to position the **bit** accurately. When the **bit** is inserted, turn the crank handle the other way, still holding on to the chuck until the **bit** is secure.

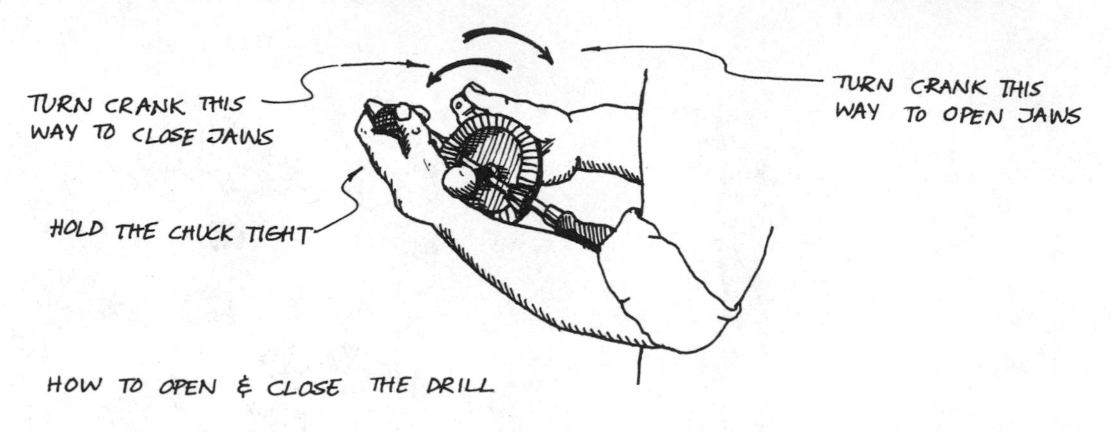

HOW TO OPEN & CLOSE THE DRILL

To remove the **drill bit**, reverse this procedure.

When starting to bore, make a small hole first with a **brad awl**. This guides the **drill bit** and prevents it from wandering. The most important point to remember is to hold the **drill** straight and do not wobble, otherwise the hole you bore will be oversize and out of line, and you may break the **drill bit**.

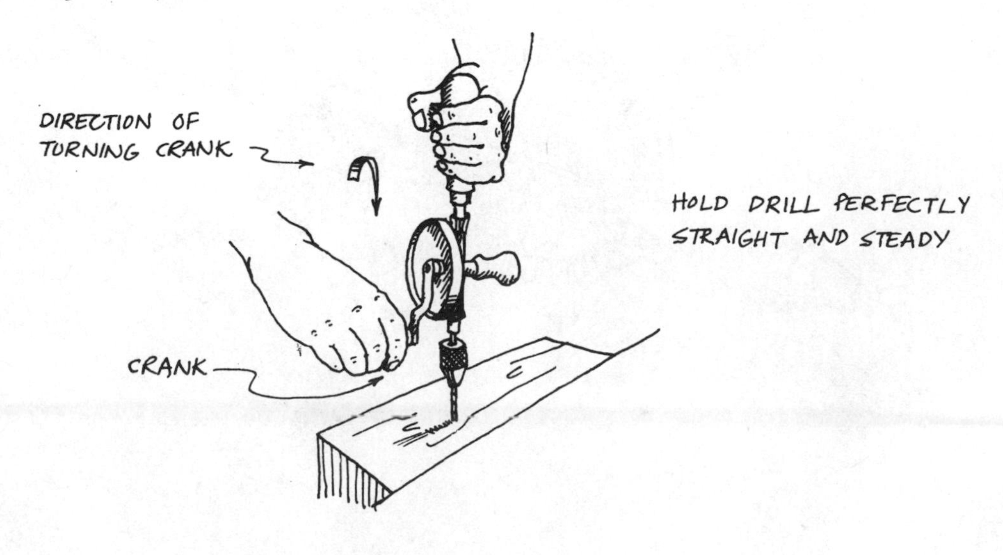

Push with an even pressure and turn the crank at a constant speed: turning too fast or too slow may break the **drill bit**, just as will too heavy a pressure.

The second most important rule to remember is that when the hole has been bored to the depth required, and you now want to remove the **drill**, under no circumstances may you reverse the direction in which you have been turning the crank handle. The temptation to do this is especially strong if you find that the **drill** does not come out easily, but it is precisely in this case that you must be doubly certain not to reverse the direction of the crank, for all that will happen is that you will unwind the chuck and leave the **drill bit** embedded in the wood! What you should do is to continue to turn as if boring, but pull on the handle instead of pushing. Also, do not attempt to waggle the **drill** from side to side; you will not work the **bit** loose but break it. The best thing to do is to try to avoid this situation of being unable to remove the **drill** easily by removing it regularly and at short intervals as you bore.

To ensure that you bore to the right depth (in cases where you do not want to bore all the way through the wood), measure the depth of the required boring against the **drill bit** and wrap a bit of sticky tape around it to serve as a marker.

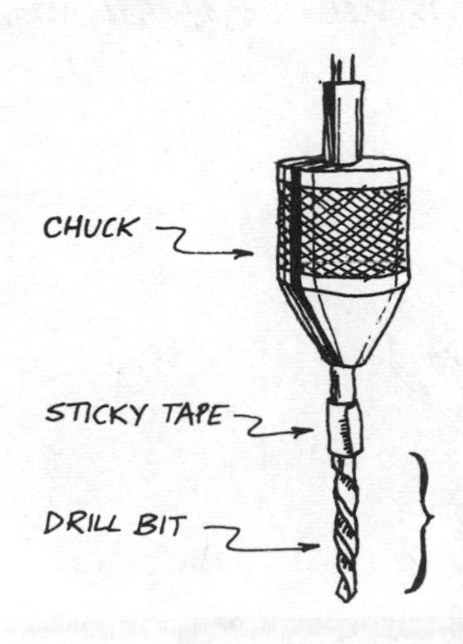

A SIMPLE DEPTH GAUGE

Sometimes it is more convenient to hold a **hand drill** by the side handle and lean against the end handle, in the same manner as a **breast drill** is used. The **breast drill**, while being constructed on the same principles as a **hand drill**, is intended for boring much larger holes.

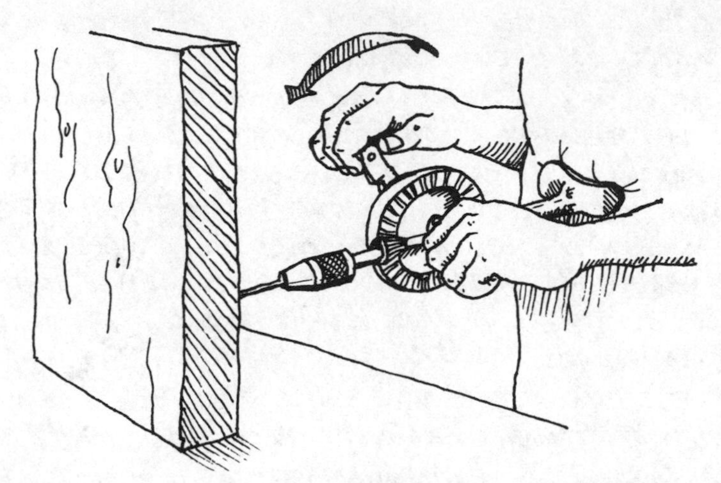

HOW TO HOLD A BREAST DRILL

The **push drill** is used for boring small holes in wood, and although it is slower than using a **hand drill** it has the advantage that it may be operated with one hand, leaving the other hand free to hold the work. The handle is usually of the magazine type, storing the **drill bits**.

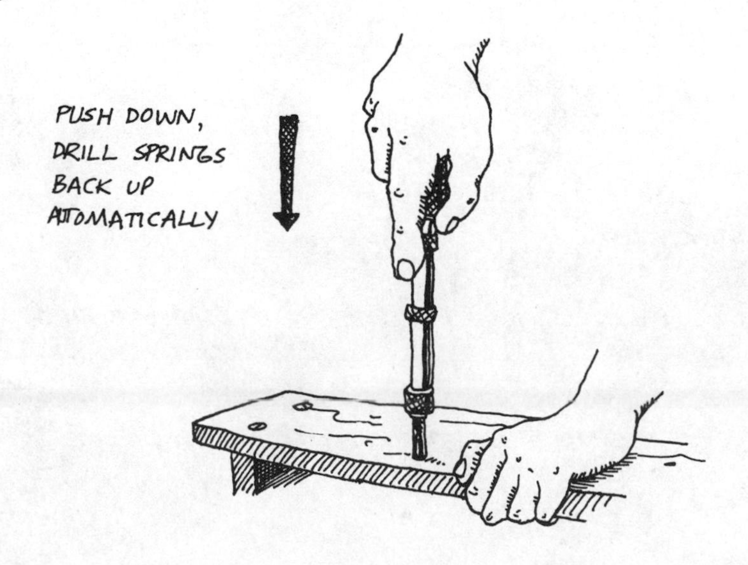

HOW TO USE AUGER BITS & BIT BRACES

The **auger bit** is fixed in the **bit brace** in much the same way as the **drill bit** is fixed in the **hand drill**. The chuck is held firmly and the handle of the **brace** is turned anti-clockwise until the jaws of the chuck are opened sufficiently wide to admit the end of the **bit**, when the chuck is then tightened again.

Care must be taken to seat the square tang of the **bit** in the V-grooves of the jaws.

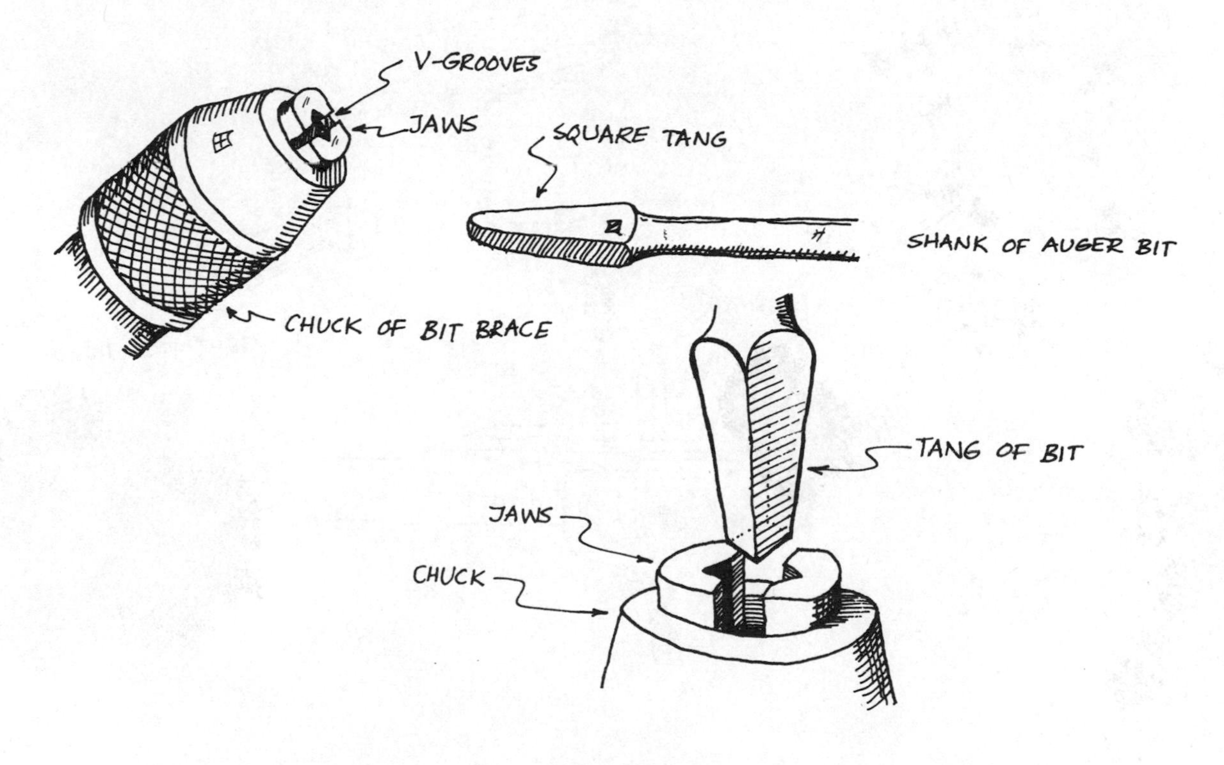

To start boring, guide the feed screw of the **bit** to the center of the hole, which should be started already with a **brad awl**.

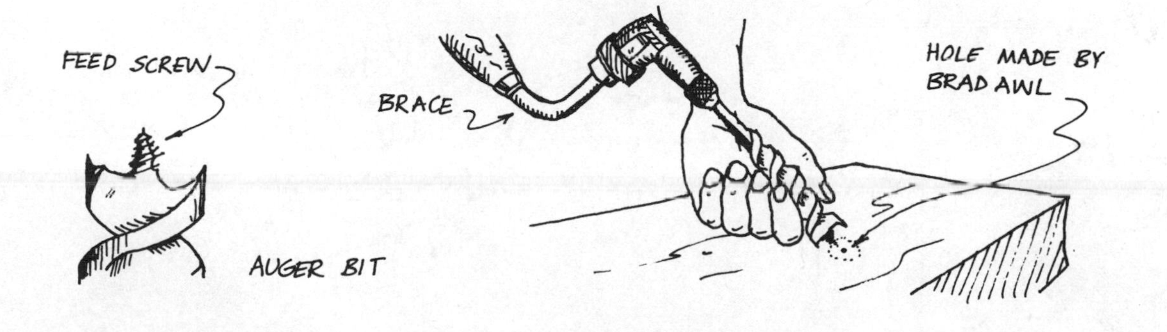

Once the **auger bit** starts boring, your main concern is to keep it perfectly perpendicular to the surface; unless you are deliberately boring a hole at an angle. You can check this by placing a **square** on the work, as shown below, but it is better to learn to develop your eye. Stop every once in a while and, holding the **brace** steady, step back a bit and sight the angle, and then, without moving the **brace**, move around to another side and sight again. What appears straight from the front may be off from the side.

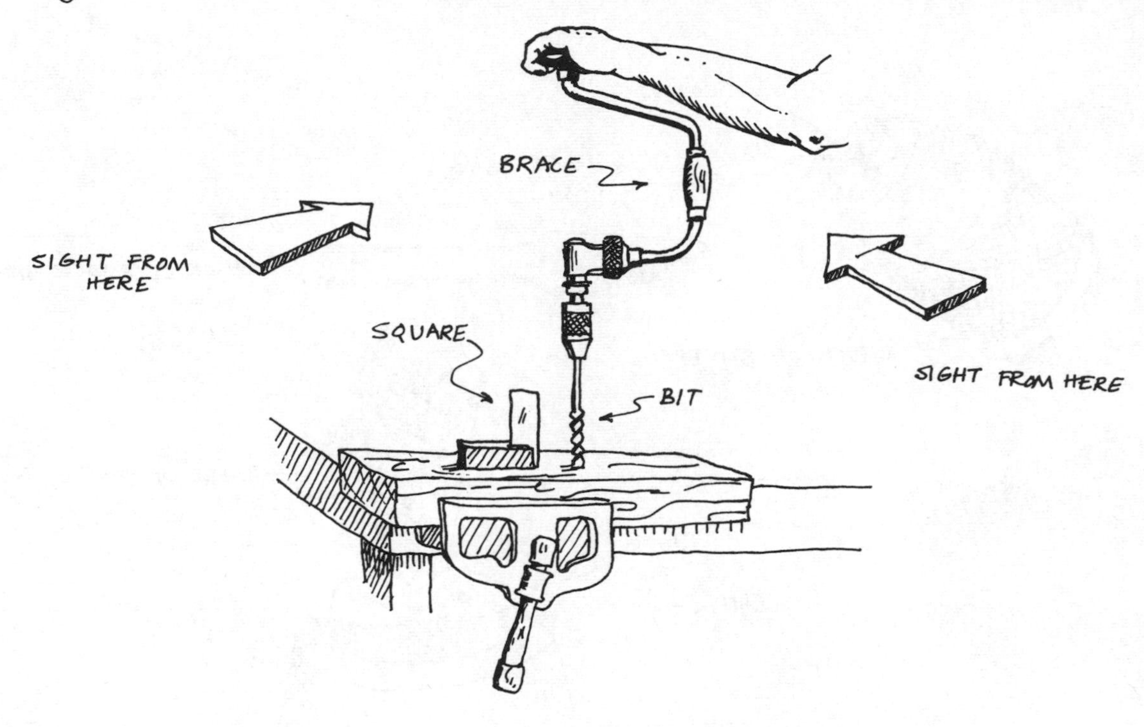

When boring a horizontal hole, hold the button, or head, against the body and turn the handle with your right hand.

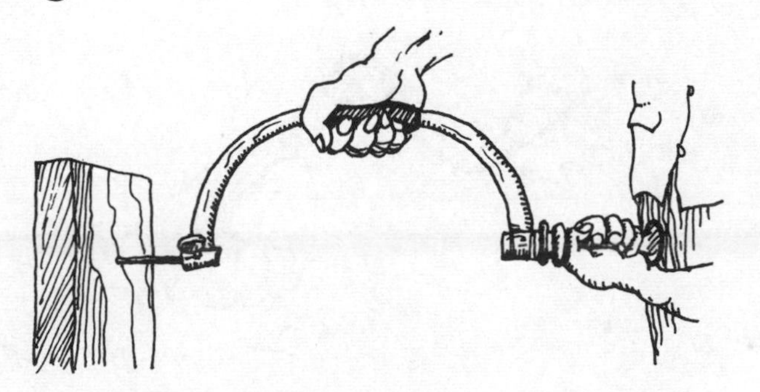

AFTER A 19TH CENTURY ILLUSTRATION OF HORIZONTAL BORING

Boring a straight hole is something that may only be attained with care and practice, but the other major problem encountered in boring, namely splitting, may be eliminated by either of the following techniques:

Boring large holes in narrow pieces of wood sometimes causes splitting as a result of the wedging action of the feed screw. Prevent this by clamping the piece tightly in a **vise** and boring a small pilot hole first, with a **hand drill**.

When boring all the way through a piece of wood, stop as soon as you can feel the feed screw emerge on the other side and turn the wood around and start from the other side.

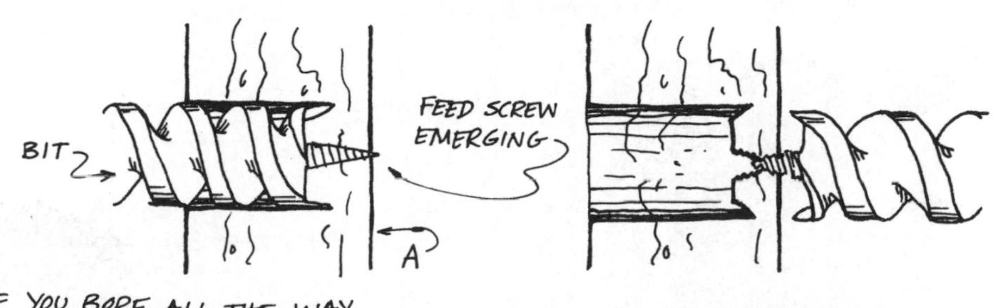

An alternative method is to clamp a piece of scrap wood behind the work and bore into that.

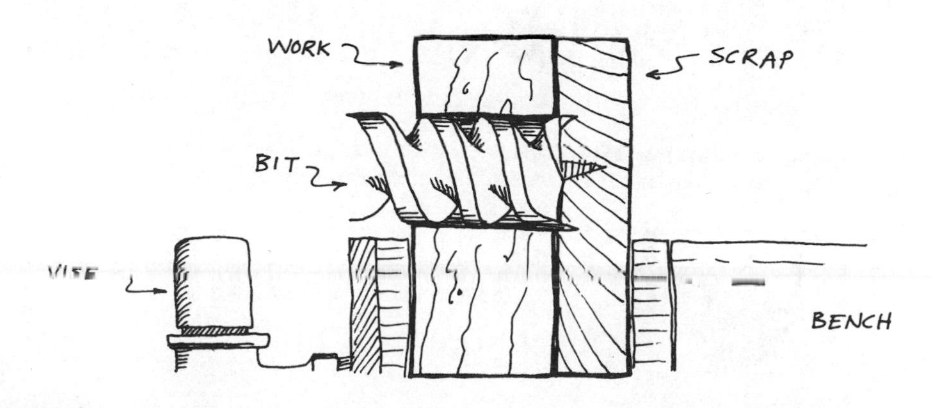

TACKS AND NAILS.

CUT TACKS, SHOE NAILS, WIRE NAILS,

Pat. Brads, Finishing Nails, Clout Nails, Trunk Nails, Hungarian Nails, Cigar-Box Nails, Basket Nails, 2d and 3d Fine Nails.

Carpet Tacks, Upholsterers' Tacks, Gimp and Lace Tacks, Brush Tacks, Copper and Brass Tacks,

BRASS AND IRON ESCUTCHEON PINS, &c., &c.,

MANUFACTURED BY

DUNBAR, HOBART & WHIDDEN, So. Abington Station, Mass.

New York Salesroom, 39 Warren St. Goods made to order from sample.

☞ Particular attention given to orders for EXPORT. ☜

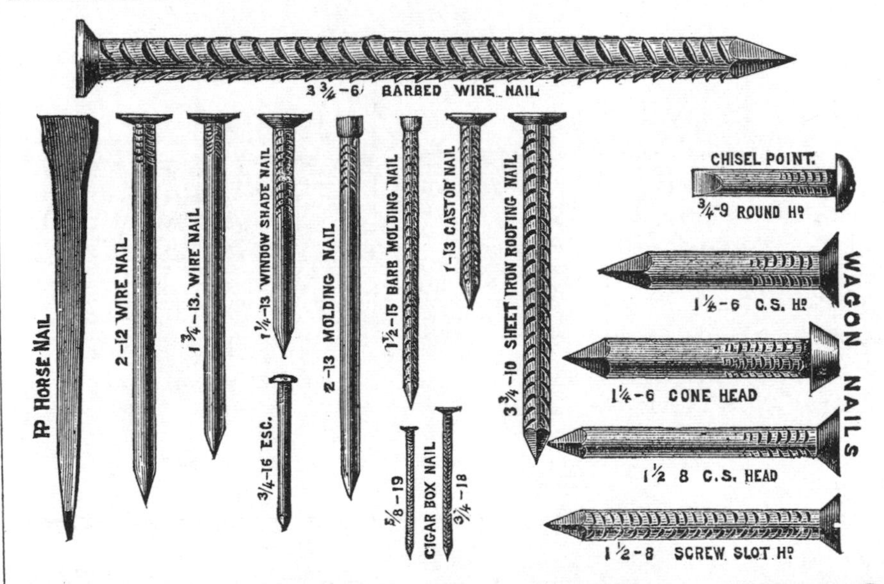

HORSE SHOE & WIRE NAILS

Steel, Iron and Brass Nails and Barbed Nails

Of every kind.

Roofing and Moulding Nails, Escutcheon Pins, Chair and Caster Nails, Cigar Box and Window Shade Nails, Wagon and Boat Nails.

Manufactured by

THE HP NAIL COMPANY,

Cleveland, Ohio.

CHAPTER SIX

Striking ~

ALWAYS MAKE SURE THE HANDLE AND THE TOOL ARE PROPERLY CONNECTED !

STRIKING TOOLS There are a great many striking tools, all of which act by blows, or by being used to hit something. They range from tiny **jeweler's hammers** to the very large and heavy tool called a **commander**.

Hammers are the commonest striking tools, and there is a different **hammer** for almost every trade; everyone, from a shoemaker to a bookbinder, has his own kind of **hammer**. The commonest woodworking **hammers** are the **tack hammer**, used for light work, such as driving small tacks; the **claw hammer**, which, while being the standard, regular **hammer**, is made in a variety of styles, sizes, and weights; and the heavy **framing hammer**, used on construction sites.

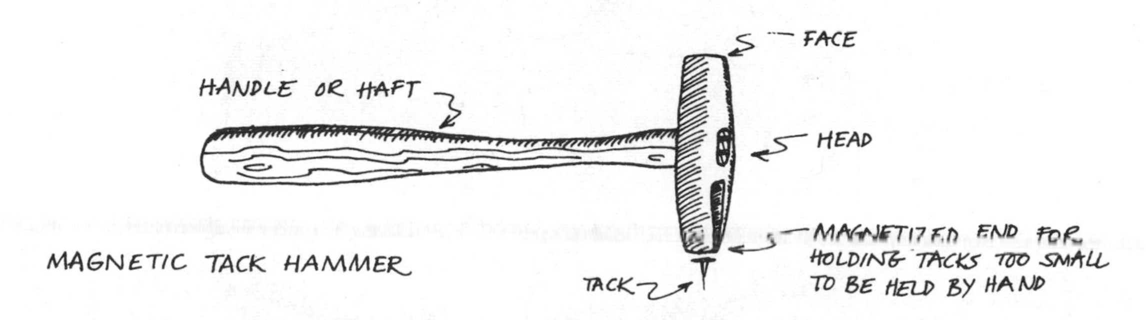

MAGNETIC TACK HAMMER

Other striking tools are the **mallet**, which is used for hitting wood; the **beetle**, which is a much heavier, two-handed **mallet**; a **maul**, which is a short-handled, lightweight **sledgehammer**; the **sledgehammer** itself, the commonest, large hitting tool; and the **commander**, the largest of all woodworking tools, used for knocking heavy beams and posts into place.

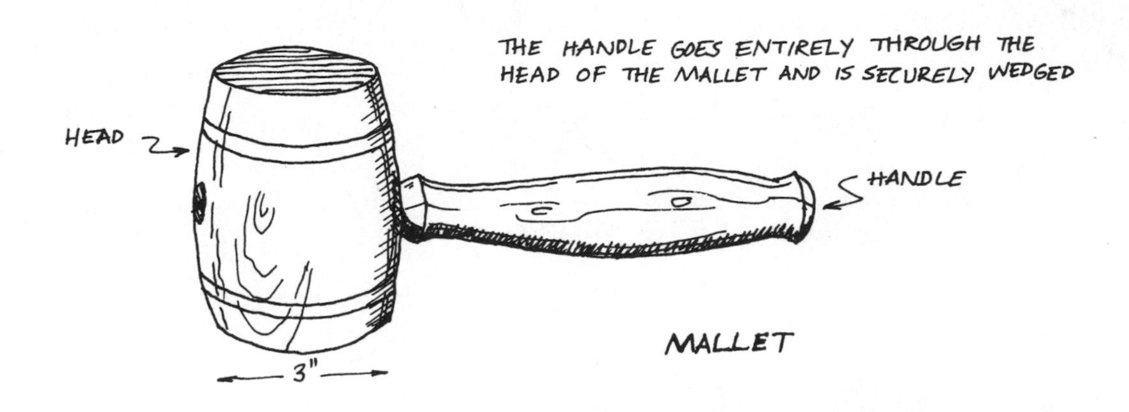

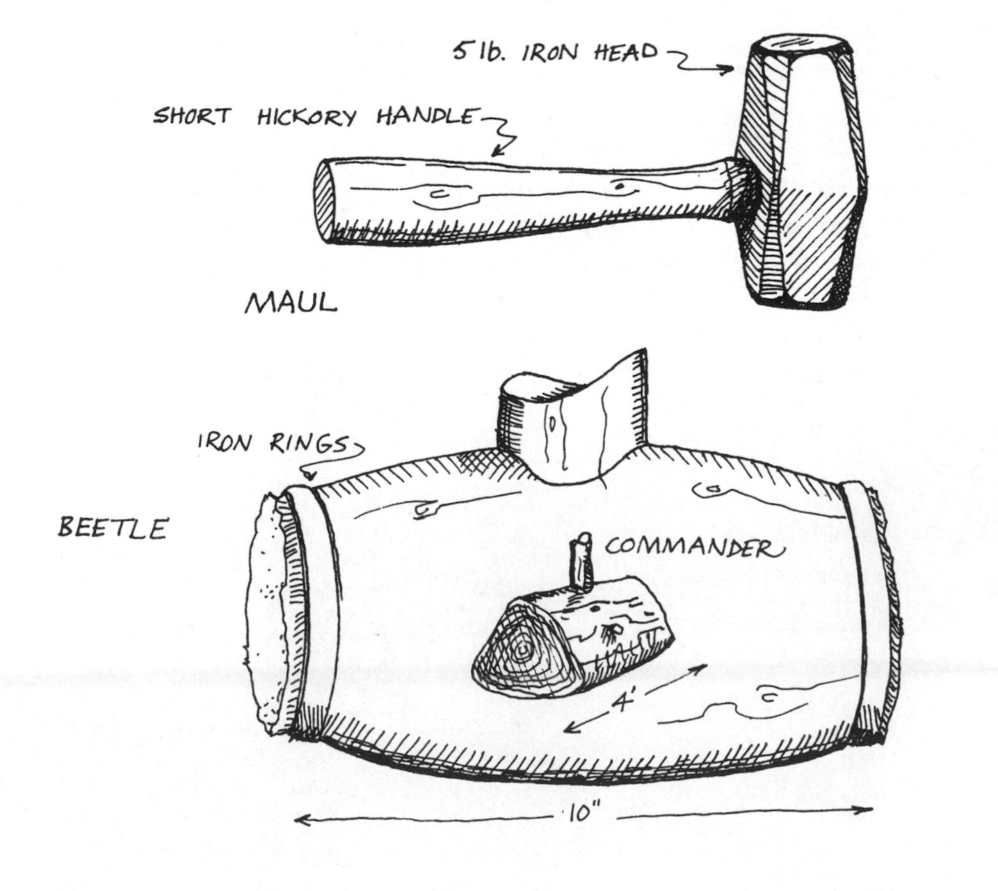

For simple woodwork all you will need, however, is a **hammer** for driving nails, and a **mallet** for hitting wood — there is a simple but important rule regarding when to use a **hammer** and when to use a **mallet** : wood to wood and metal to metal.

Choose a **hammer** that feels comfortable, bearing in mind that the heavier the **hammer** the fewer the strokes you will need, but the quicker you will tire. Remember too that it is difficult to drive big nails with anything much lighter than a 10 oz. **hammer**, while, even though you may feel very strong, small nails and tacks are hard to hit with 20 oz. **hammers** —there is simply too much **hammer** to even see the tack!

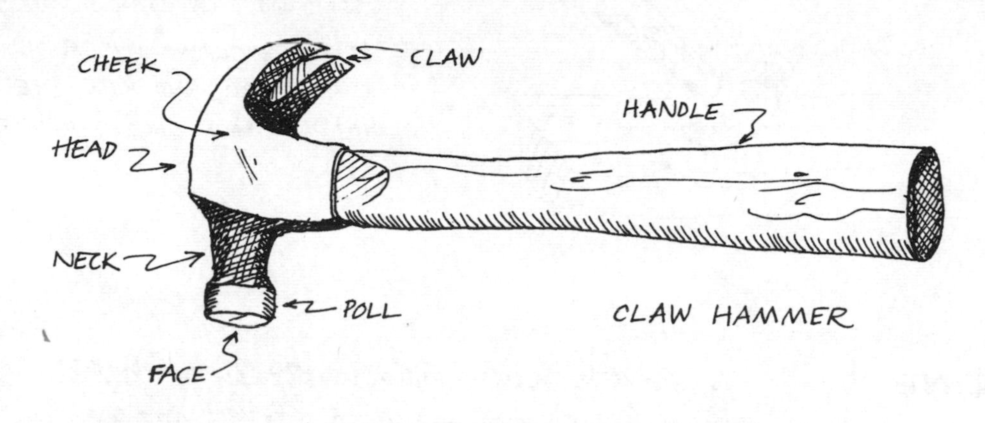

There are two kinds of faces on **hammers** : bell-face, and flat or plain-face. Bell-face **hammers** have striking surfaces that are convex, and can drive nails flush with the surface of a board, or even slightly below, without leaving hammer-marks. Plain-face **hammers** can't do this, but, having flat faces, are easier to learn to use.

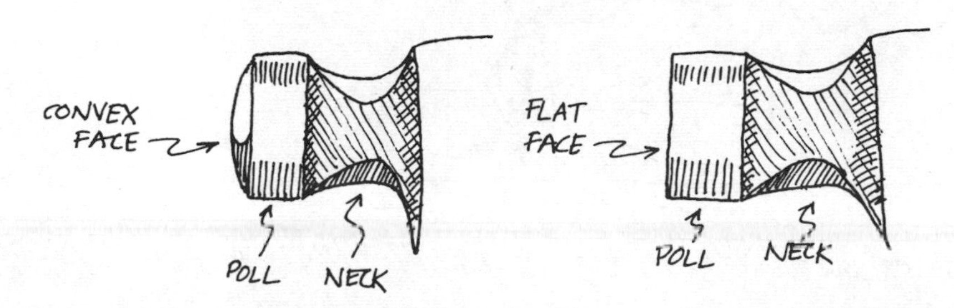

BELL-FACE HAMMER PLAIN-FACE HAMMER

Claws on **hammers** also vary in shape to facilitate removing nails. While it is all right to remove small nails with the claw, the practice should be discouraged since there exists the temptation to pull ever larger nails and turn the **hammer** into a wrench, which weakens the hold of the handle in the head and may spoil or break the handle.

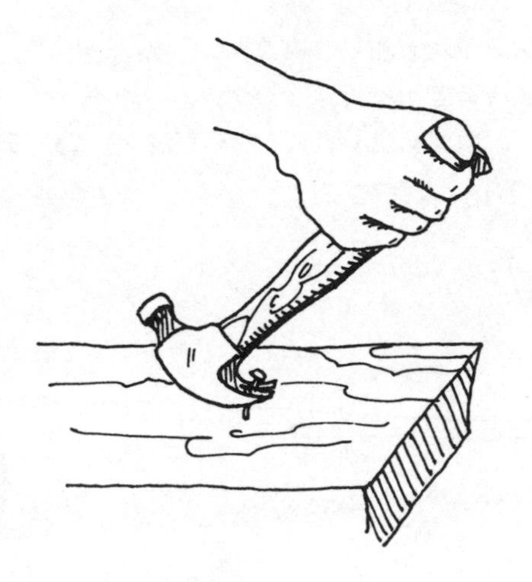

TO REMOVE A SMALL NAIL: SLIP THE CLAWS UNDER THE NAIL HEAD AND PULL THE HANDLE BACK LIKE A LEVER

HAMMERING To start a nail, hold it steady with the thumb and fingers and strike one or two light blows with the **hammer** until the nail is well started.

HOLD THE HANDLE NEAR THE END.
MAKE SURE THE FACE IS CLEAN OF GREASE OR DIRT

After the nail is started, drive it in with firm, well-directed blows. In order not to bend the nail, it must be hit squarely on the head and not obliquely.

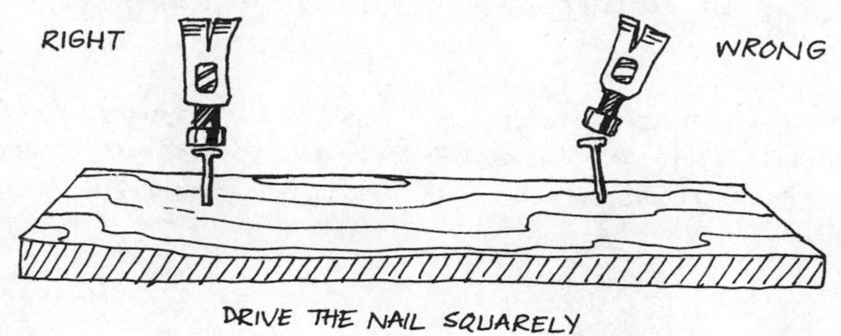

DRIVE THE NAIL SQUARELY

Try always to strike with precision, otherwise you will end up leaving "moons" in the wood.

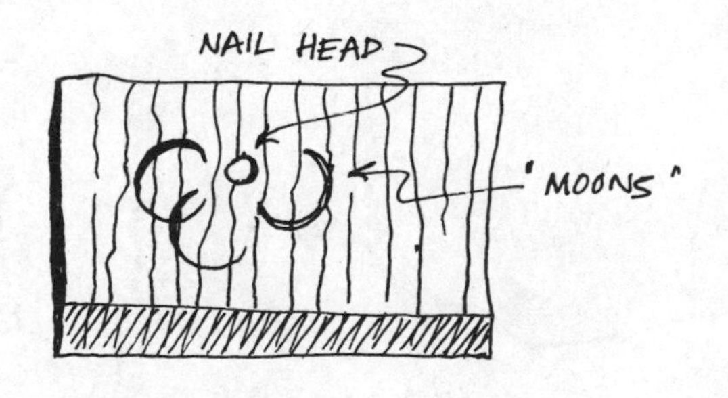

Mark a number of points on a board and hit them with the **hammer** as squarely and as precisely as possible. Before swinging the tool, place it on the point you wish to hit and then try to bring the tool, and all parts of the body, into exactly the same positions, at the instant the blow occurs, as they had when the tool was placed in position.

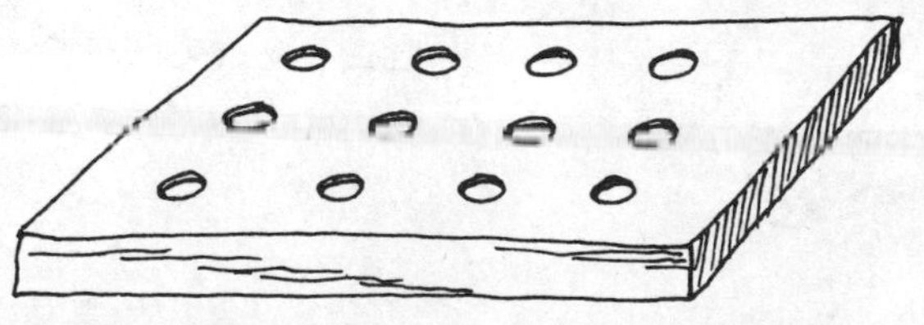

THE EFFECTS OF SQUARE BLOWS

Sink the face of the **hammer** many times into the surface of the wood. Try to have the centers of the dents about an inch apart, and on parallel lines, as shown on the previous page. Be very particular to strike square blows. The dent made by a square blow will have the same outline as the face of the **hammer**, and all points of its outline will be sunken equally deep.

For driving small nails, use mostly wrist movement; for heavier hammering use wrist movement and elbow movement; and for very heavy hammering, use shoulder movement as well.

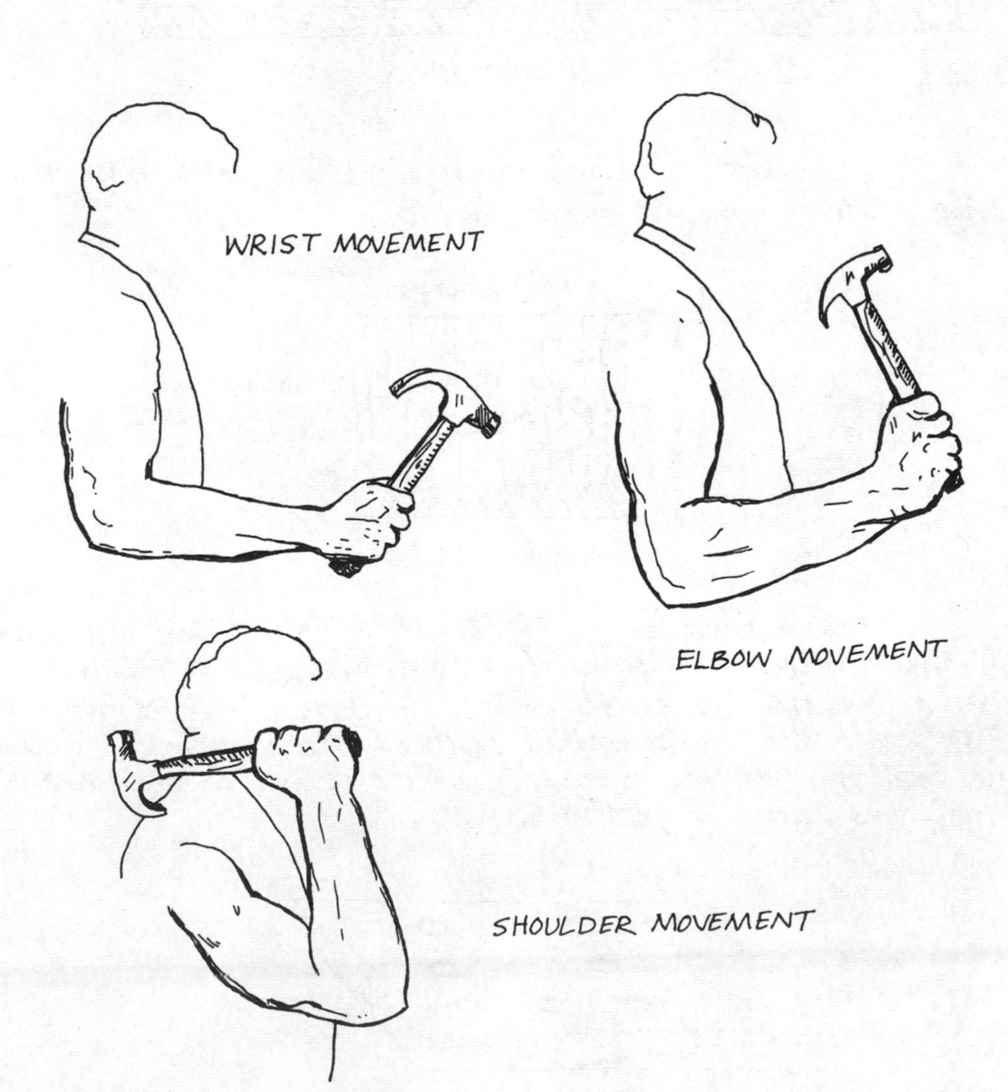

If a nail is placed too near the end of a board it will split the wood. Avoid this by using a smaller nail, by pre-drilling, or by hitting the nail a couple of times upside down. This last action blunts the end of the nail and creates a small depression in the wood. When the nail is reversed and driven into this depression, there is less chance of the wood splitting.

Try not to nail in a straight line along a board, but rather in some kind of pattern. This is to avoid having too many nails enter the same line of grain, which might cause the wood to split.

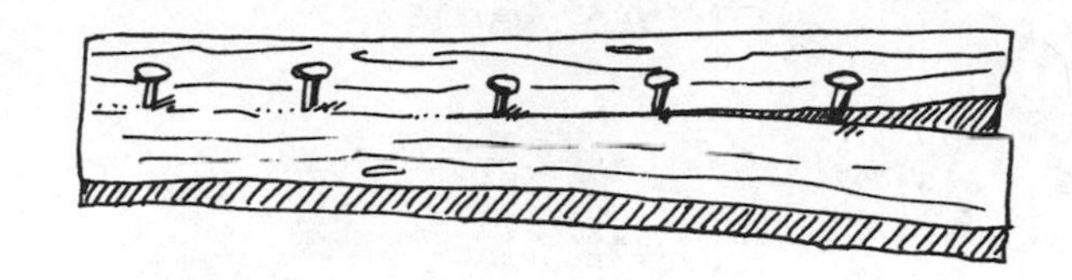

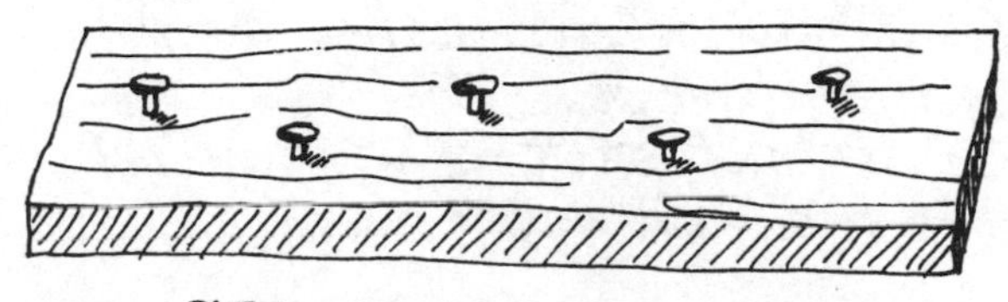

PATTERN OR STAGGER NAILING

To start a nail that can't be reached by both hands, hold the hammer by the head, with the nail against the cheek. This will start the nail which can then be driven in the usual way.

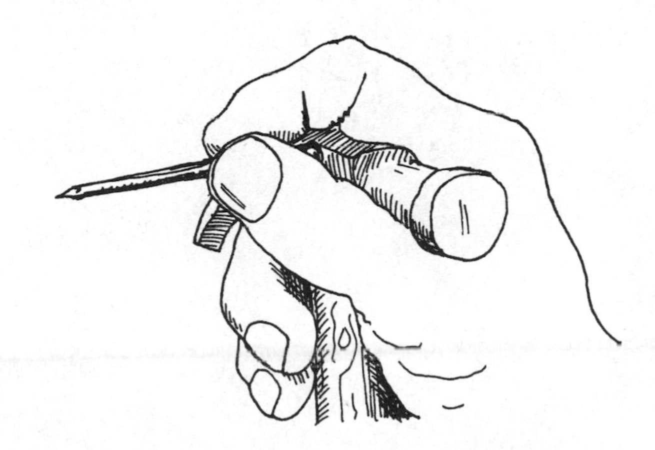

NAILING WITH ONE HAND

There are two other methods of nailing that should be practised. They are clinching and toenailing.

Nails are clinched by bending over the protruding ends. Clinching makes nails hold better, and is sometimes done for

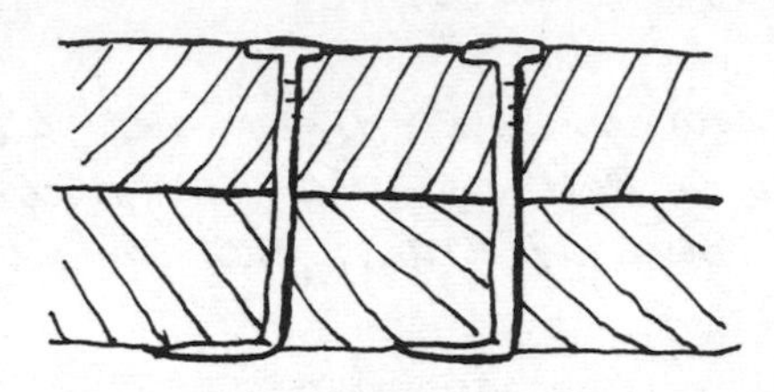

CLINCHED NAILS

decorative reasons. Many old doors were made by nailing two layers of boards together with clinched nails.

Toenailing is done to fasten a piece that butts against another. You must be careful when choosing the point to start the nail and the angle to drive it. The nail should get a good hold in the first piece without splitting it, and yet go deep enough into the second piece to ensure a good hold.

TOENAILING

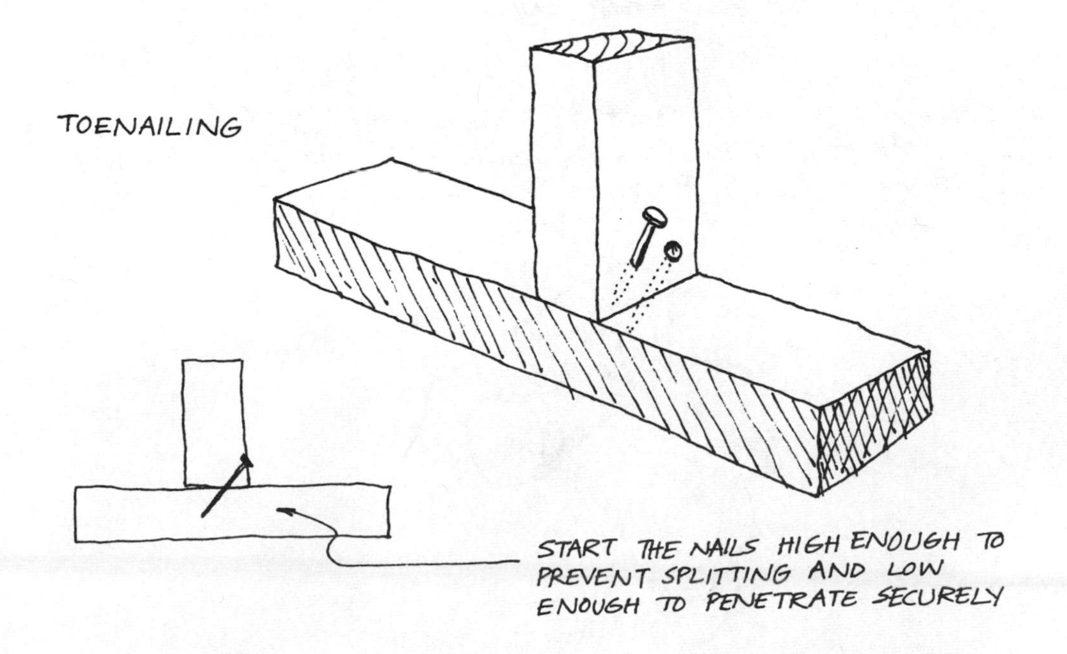

START THE NAILS HIGH ENOUGH TO PREVENT SPLITTING AND LOW ENOUGH TO PENETRATE SECURELY

The **nail set** is a tool used to set the heads of nails below the surface of the wood without marring the wood. Different sized **nail sets** should be used for different sized nails, but in any event the tip of the **nail set** should be smaller than the head of the nail it is used on.

The tip of the **nail set** is concave so that its rim can dig into the nail and not slip off when hit. The **nail set** should be placed squarely on the head of the nail with the body of the **nail set** at the same angle at which the nail was driven into the wood.

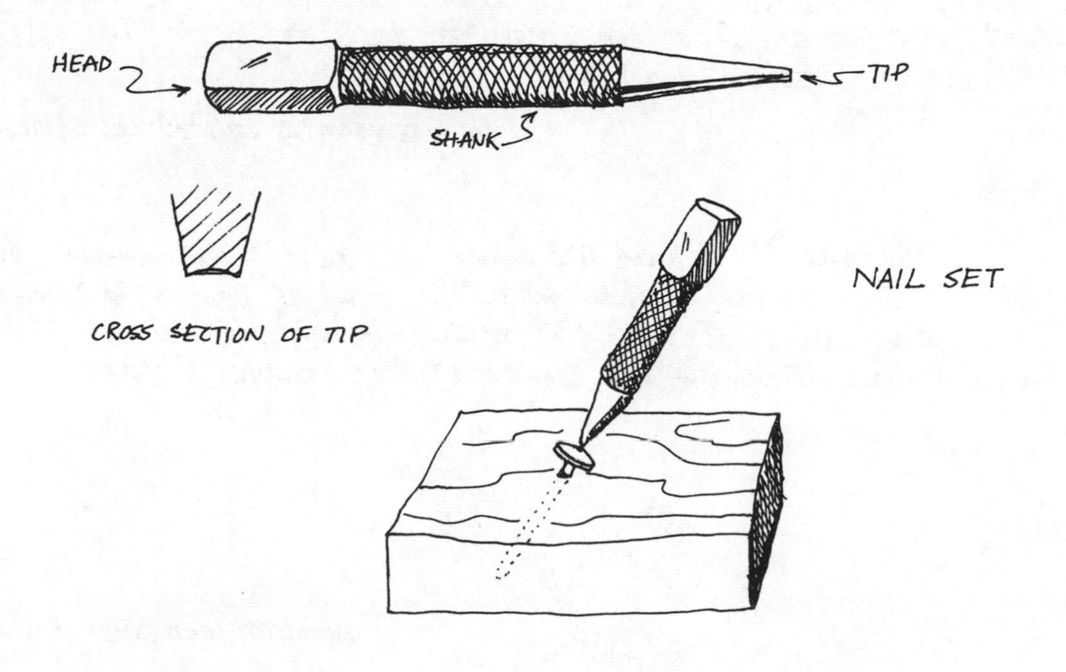

NAIL SET

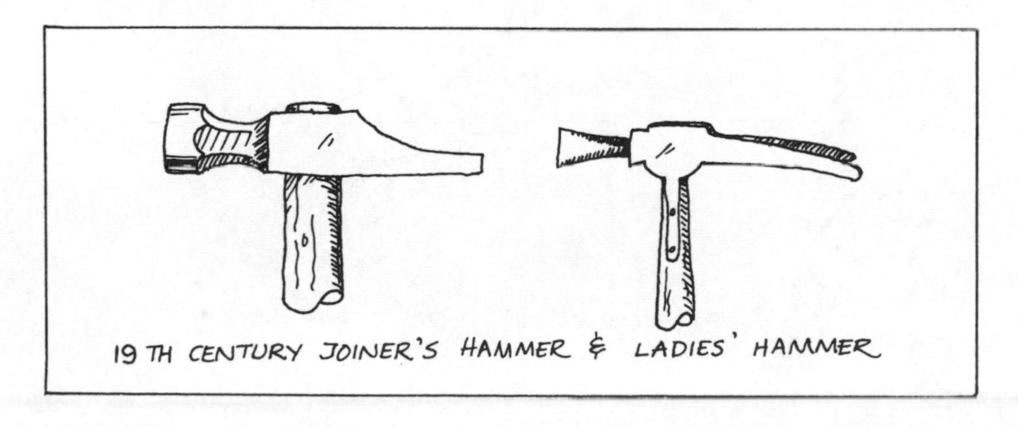

19 TH CENTURY JOINER'S HAMMER & LADIES' HAMMER

USING MALLETS

"It may seem somewhat superfluous to remark that **hammers** are meant for driving nails, striking **nail sets**, etc., and not for hitting wood; but it unfortunately happens that the amateur, and the artisan too sometimes, is given to use the **hammer** for striking the handle of his **chisel** when mortising, or the **screwdriver** in getting out obstinate nails, much to the detriment of the handle, which is bruised and split by the blows of the **hammer** and thereby rendered unfit to be held in the hand for cutting, in the case of the **chisel**, or for inserting or withdrawing screws in the case of the **screwdriver**. Wood must in all cases be struck by wood; and when it is necessary to strike the handle of a **chisel** in mortising, or the handle of a **screwdriver**, it should be done with the tool proper for the purpose, which is the wooden **mallet**."

from *Hammers & Mallets: Their Uses*

The **mallet** is held the same way as the **hammer**, but a little higher up the handle, since the head is larger and heavier. When hitting any tool with the **mallet**, be careful to hit the handle of the tool with the center of the **mallet's** face.

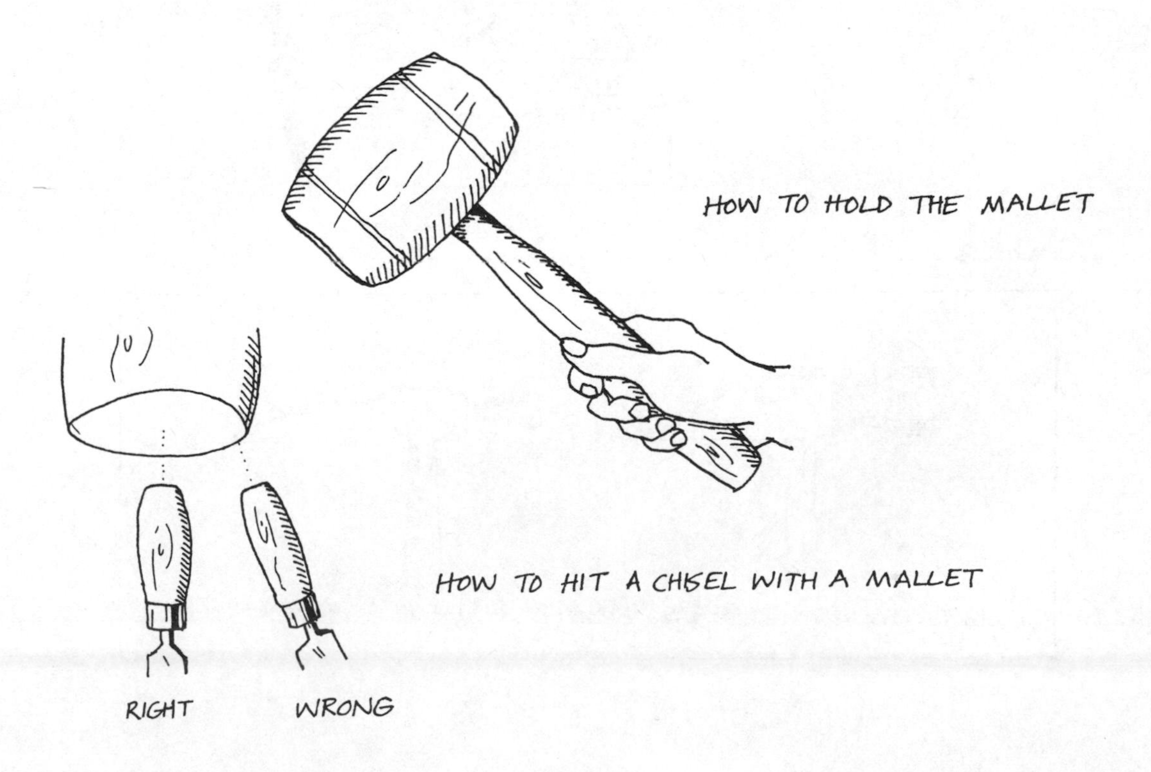

TIGHTENING HANDLES Handles of **hammers** and **mallets** often come loose, and consequently become very dangerous; they should be retightened at once. This is usually very easily done, all that is needed is a **mallet** and some small steel wedges, which can be obtained for the purpose from most hardware stores.

The handle is driven into the head either by being hit on the end with a **mallet** or by being pounded onto some hard surface.

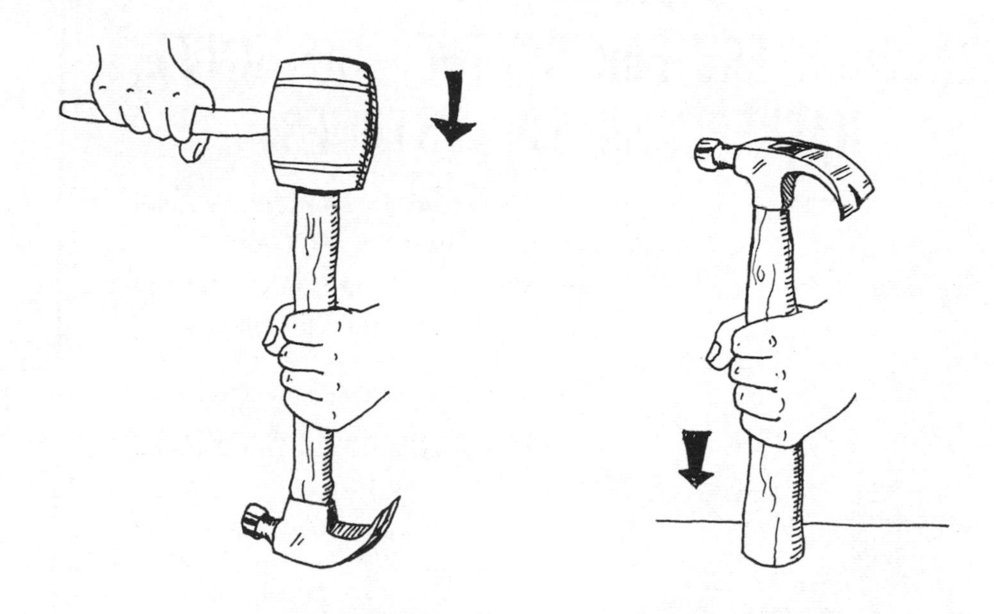

When the handle is securely into the head, wedges are driven into the "eye" of the tool with a **hammer**.

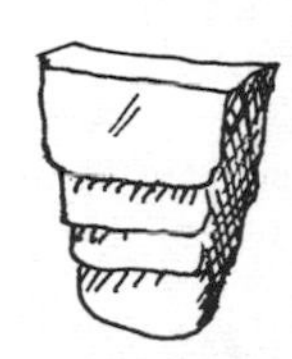

STEEL WEDGE, ACTUAL SIZE

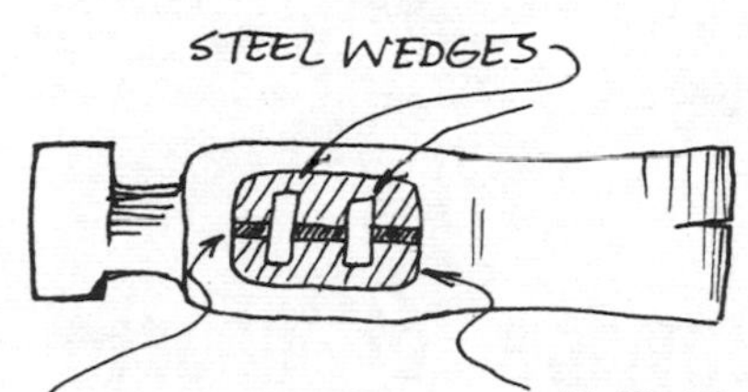

A. F. PIKE,

Pike Station, NEW HAMPSHIRE.

(ESTABLISHED 1823.)

HEADQUARTERS FOR SCYTHE, AXE, KNIFE, HACKER AND TOOL STONES.

Twenty Quarries and Four Factories in New Hampshire and Vermont.

Strong, Clear Grit Stone that will not glaze.

PRICES & QUALITY GUARANTEED

All Goods Genuine Brands.

My customers may rely upon being squarely dealt with and getting no poor, unsalable imitations.

LIST

No. 1, Extra Indian Pond.
No. 1, " "
No. 2, " "
Premium.
Union.
White Mountain.
L'étoile.
Diamond Grit.
Hacker (Round).
Lamoille.
Willoughby Lake.
Green Mountain.
Black Diamond.
Ragg.
Mowing Machine.
Paper Mill Stone.
Vermont Darby.
" Chocolate.
" Axe Bitts.
N. H. Chocolate.
German Pattern.

The only Manufacturer of Genuine, Old Reliable Indian Pond (Red End).

Stones manufactured, labeled and branded in any manner desired.

Beware of Coarse Brittle Imitations.

CHAPTER SEVEN

Tool Care & Sharpening –

SHARPENING SAVES MORE TIME THAN IT TAKES !

It does not matter how carefully you work, or how good the quality of your tools is, you will be unable to do anything properly, far less well, if you do not care for your tools adequately. This means they must be clean, free from rust, tight where necessary, and always sharp.

A professional woodworker will frequently stop in his work to put his plane iron or chisel to his sharpening stone, for while the inexperienced worker is often too impatient to take the time, the successful craftsman knows that blunt tools cost far more time than it takes to sharpen them.

After some general observations on care and storage, this chapter explains how to maintain and sharpen the tools described so far, and in the same order, that is:

1. SAWS
2. PLANES
3. CUTTING TOOLS
4. BORING TOOLS
5. STRIKING TOOLS

GENERAL CARE

The more tools you have, the more tools you will use. It is sometimes quite astounding how many tools will be used for an apparently simple job. Therefore, in order to be able to work with any efficiency, it is imperative that you be tidy. This means you should put a tool away as soon as you have used it. Your work will be much easier if you are able to remember the old maxim: "**A place for everything and everything in its place!**"

1. Get into the habit of folding up your **rules**, **tape measures**, and **bevels** after use.

2. Protect your edges! A **rule** is not much good for drawing a straight line if it is nicked. Nor do **chisels** or **knives** cut very smoothly if their cutting edges look like **saws**; and **saws** themselves must have sharp teeth if they are to cut easily. When you lay your **chisels** down make sure the beveled side is down so that the cutting edge is not touching anything.

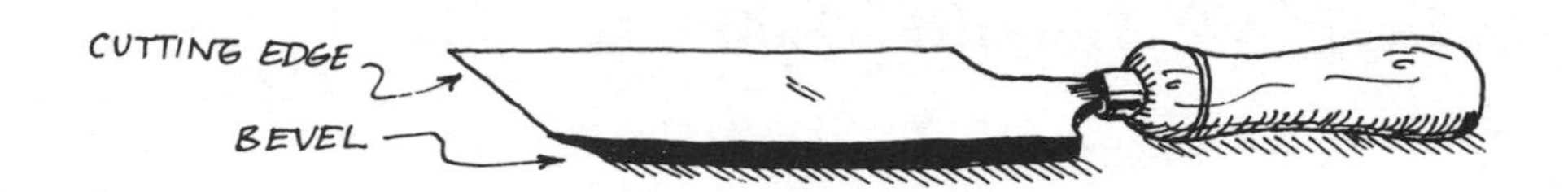

   Similarly, take care to lay your **planes** down on their sides, not on their bottoms, where the cutting edge would touch the bench, and after finishing with a **plane** retract the blade so that it no longer protrudes.

3. Store things so that they are protected from each other, and at the same time are easily accessible. If you throw all your tools into a box, not only will they become blunt and damaged by banging against one another, but you will discover that whatever tool you need, it is always at the bottom of the pile!

   Hang your **saws** up so that nothing can knock into the teeth. Store your planes on a shelf. Make a rack for your **chisels** so that the edges hang free. Bore small holes in the ends of handles of tools like **hammers** and **mallets** so that they may be hung on nails.

Be especially careful with your **squares**; they are a precision instrument. Do not leave them lying around where they might get trodden on. A quick test of the accuracy of a **square** is shown below. Do it often.

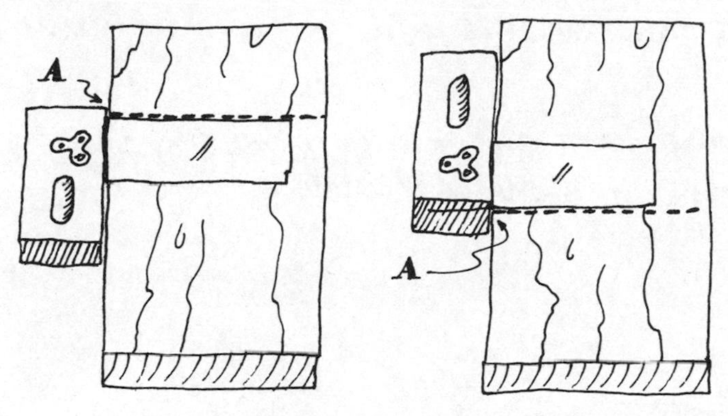

PLACE THE SQUARE ON A BOARD AND DRAW A LINE ACROSS FROM POINT **A**. TURN THE SQUARE OVER AND REPEAT. IF THE SQUARE IS TRUE THERE WILL BE ONLY ONE LINE.

IF THERE ARE TWO LINES, SQUARE IS CENTER :

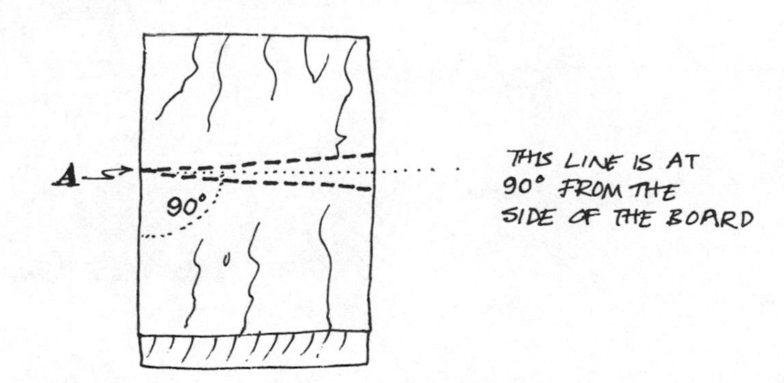

4. Protect your tools from rust by keeping them dry, and when not in use, covered with a thin film of oil or grease.

5. Oil all moving parts regularly, such as **vises**, **clamps**, **bevels**, **metal tape measures**, **hand drills**, and **bit braces**.

6. Make sure screws or nuts and bolts are tight, such as in **saw** handles or **planes**.

SHARPENING

All that is needed for successful maintenance of the tools mentioned in this book is a **grindstone**, a couple of **files**, and an **oilstone**.

Electric **grindstones** are excellent and very convenient because they can be operated at a constant speed and leave both hands free, but hand-operated **grinders** have been used for a very long time and do very well.

There are many grades and varieties of **oilstones**, but for general use a carborundum **stone** is adequate.

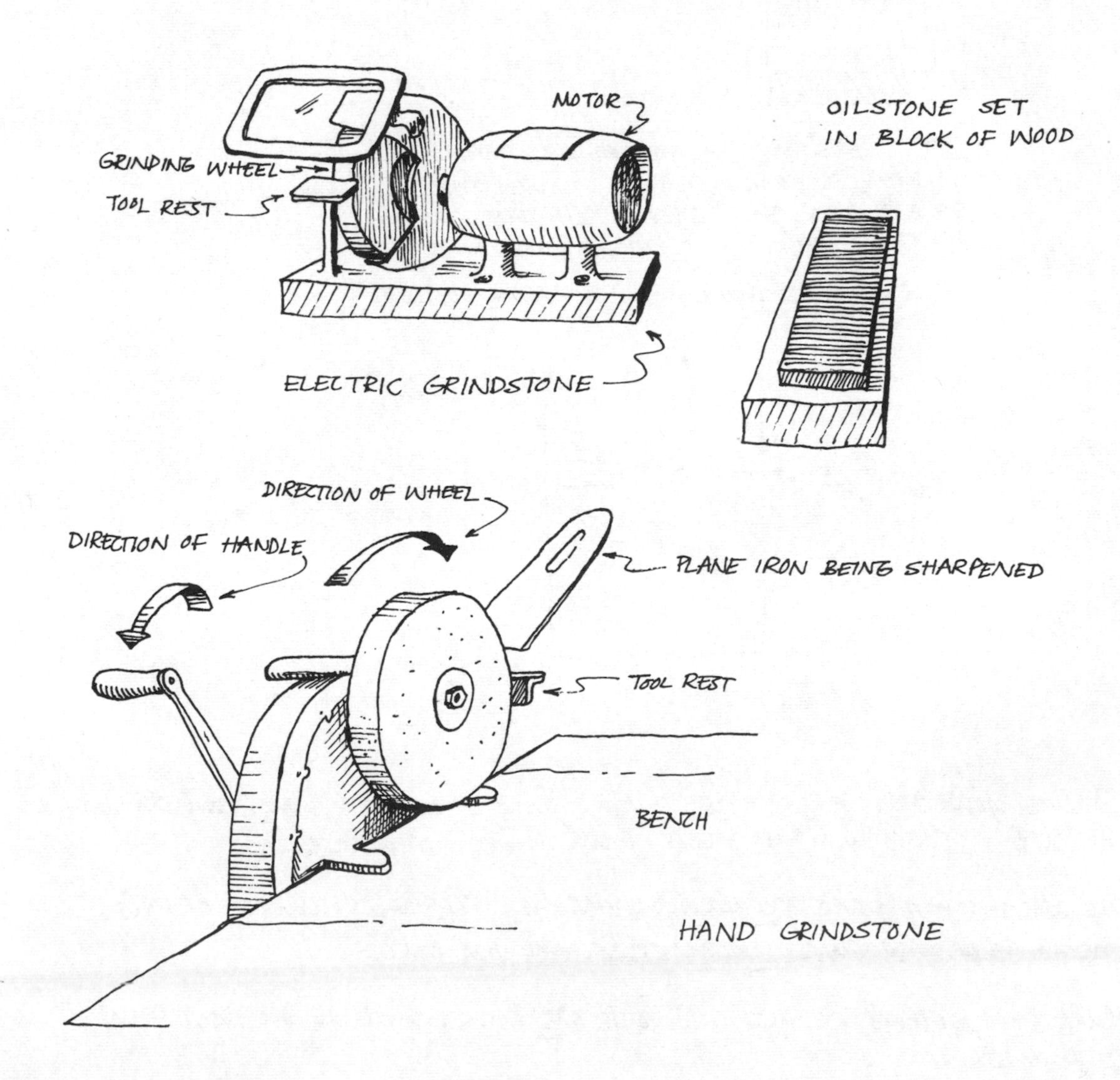

1. SAWS Learning to sharpen a **saw** is no more difficult than learning to use a **saw**, and with practice can be done adequately by anyone having sufficient mechanical ability to saw.

The three main operations are:

i. JOINTING
ii. SETTING
iii. FILING

i. JOINTING The purpose of jointing is to make all the teeth the same length. It is done very simply by running a flat **file** along the top of the teeth. Only two cautions need be observed. The first is that the teeth need be filed only until there is a <u>very</u> small shiny spot on the end of <u>every</u> tooth; the second is that the **file** must be held perfectly flat. To make this second point easier to observe, clamp the **saw**, teeth up, in a **vise**, and make a holder for the **file** as shown.

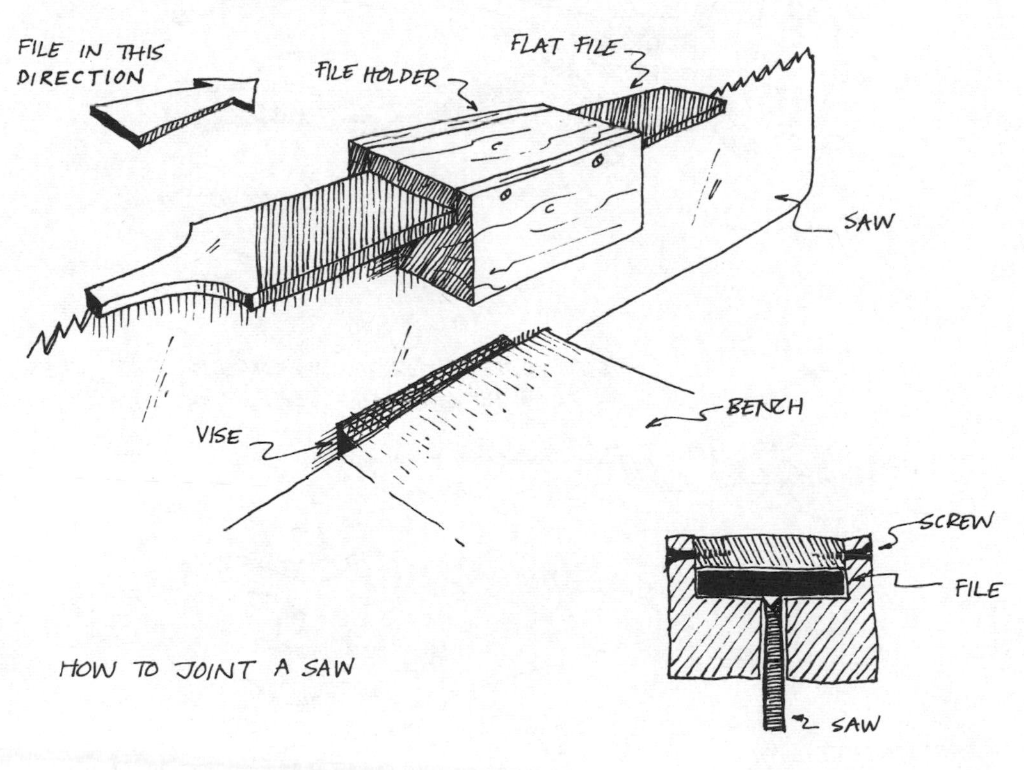

HOW TO JOINT A SAW

END VIEW OF FILE HOLDER.

ii. **SETTING** As is illustrated on page 39, the teeth of a **saw** are bent out a little in order to cut a kerf wider than the blade so that the **saw** can pass easily through the wood. Usually, the teeth need not be set every time the **saw** is sharpened, every third or fourth time is generally sufficient.

Teeth are bent out the correct amount with a tool called a **saw set**. Most **saw sets** are adjustable, so experiment first until you find the right amount of set for your **saw**. This will depend on the use to which you put the **saw**. Green or wet wood needs more set than dry or seasoned wood, and softwoods need more than hardwood. Too much set will give you a kerf which, being too wide, will make for harder sawing and less control, while too little set will cause the **saw** to bind in the kerf.

SAW SET

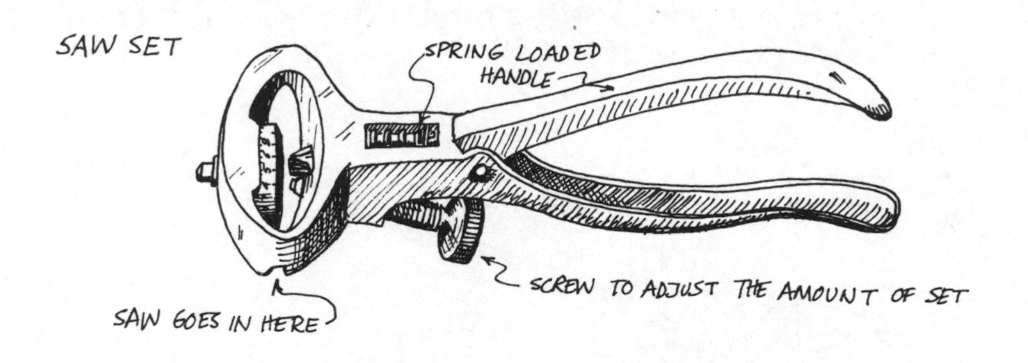

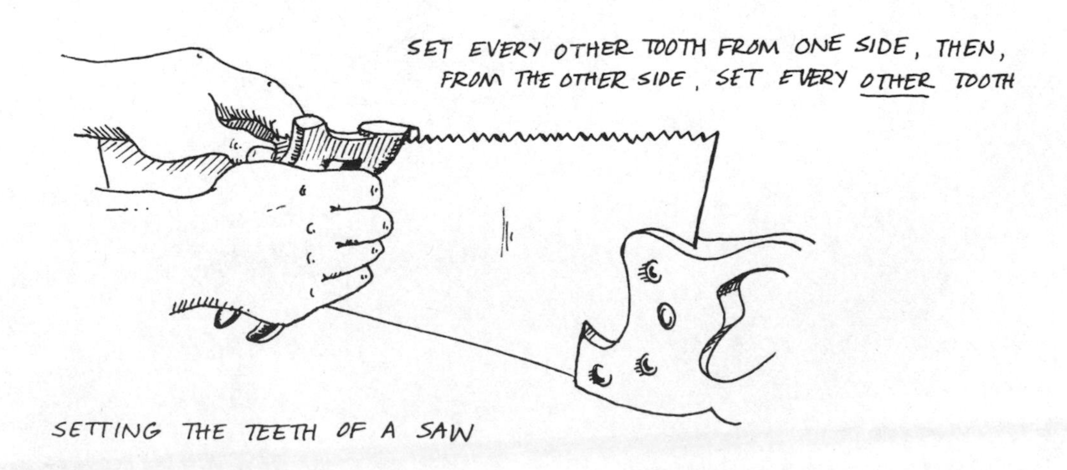

SETTING THE TEETH OF A SAW

iii. FILING Filing is the most difficult operation, and the one which requires the most practice. Before you begin to file, study very carefully the shape and cutting edges of the teeth of a new **saw** or one newly sharpened by an expert. See page 39. Notice that different types of **saws** have different kinds of teeth, which must be filed accordingly.

To have any success at all, you must clamp the **saw** really securely in a **vise** so that the teeth project no more than ¼" above the jaws of the **vise**. If you do not do this, the **saw** will chatter and vibrate, and the **file** will screech unbearably.

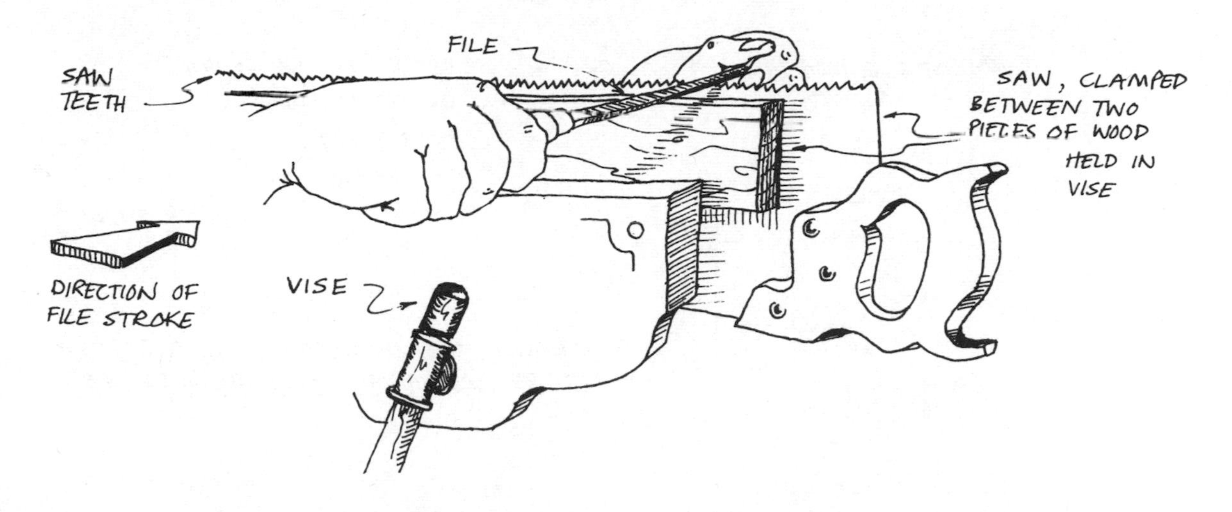

1. In general a 6" **slim-taper file** will suffice.
2. Hold the **file** level.
3. Begin at the toe of the **saw** and work towards the handle.
4. File all the teeth from one side first, then turn the **saw** around and file all the teeth that point to the other side.
5. Exert only enough pressure on the **file** to make it cut, and do so on the forward strokes only.
6. When you have filed both sides check to see if there are still any flat and shiny spots left from jointing, and if so, file only those teeth until all are sharp.

MOST IMPORTANT IS TO KEEP THE PROPER SHAPE AND ANGLE OF ALL TEETH.

2. PLANES

The sharpening of **planes** consists of grinding and whetting. Grinding need be done only if the plane iron is nicked, out of square, or incorrectly beveled.

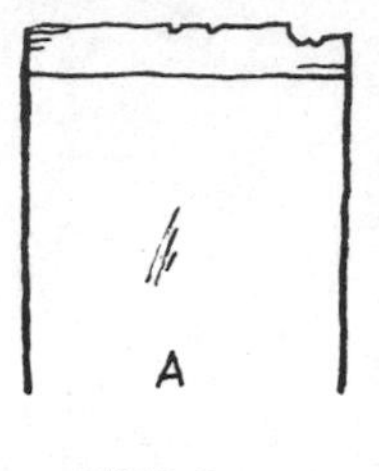

NICKED

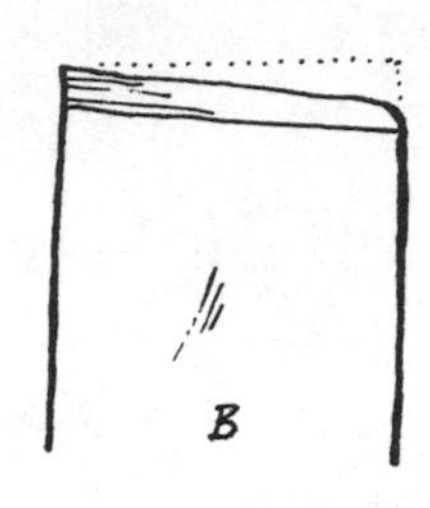

OUT OF SQUARE

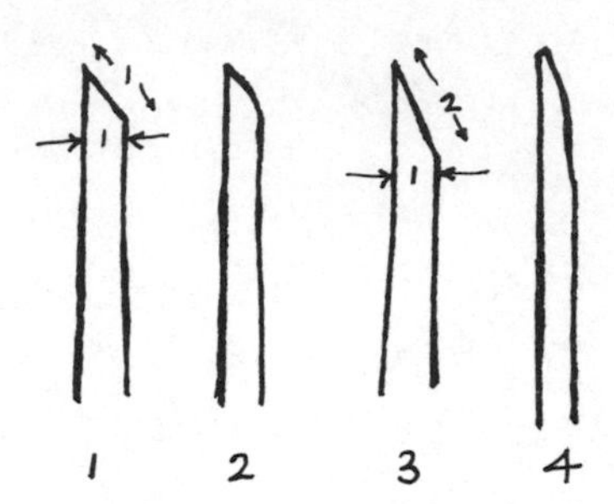

1 2 3 4

VARIOUS CONDITIONS OF PLANE IRONS

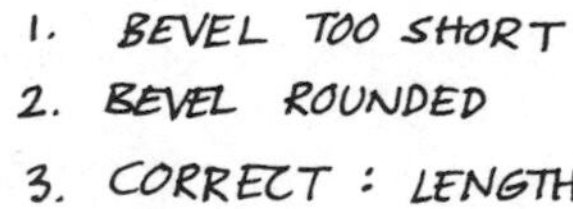

1. BEVEL TOO SHORT
2. BEVEL ROUNDED
3. CORRECT : LENGTH OF BEVEL EQUALS TWICE THICKNESS OF IRON
4. BEVEL TOO LONG & ON BOTH SIDES

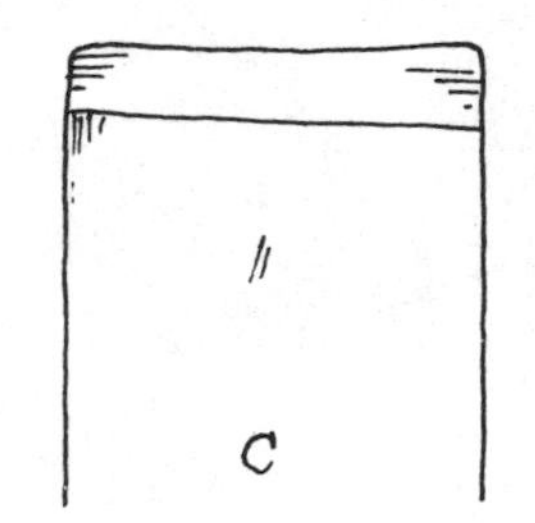

GENERALLY, THE CORNERS OF A PLANE IRON SHOULD BE GROUND <u>SLIGHTLY</u> ROUNDED, AS AT LEFT

If you must grind, remember, the angle of the bevel is between 25° and 30°. The **grindstone** should turn towards the plane iron. Prevent the iron becoming too hot by frequent quenchings in water. It is too hot when the metal turns blue; at that point, the temper of the steel has been drawn, which means the metal is now too soft to hold an edge. Check to see if you are grinding square and to the correct bevel often.

Whetting is done on an **oilstone**. The angle at which the plane iron is whetted is a little greater (30° to 35°) than that at which it is ground (25° to 30°).

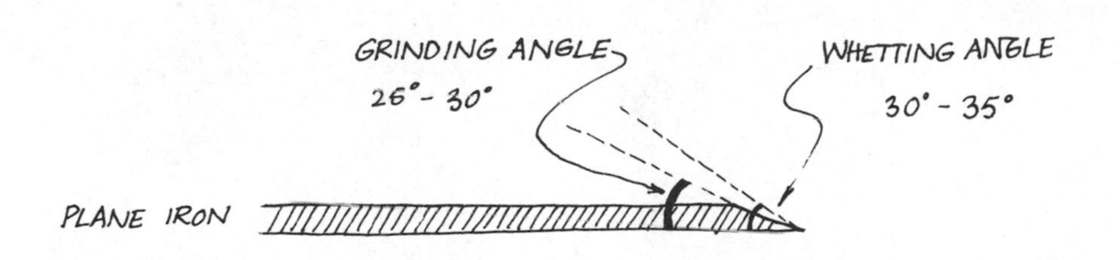

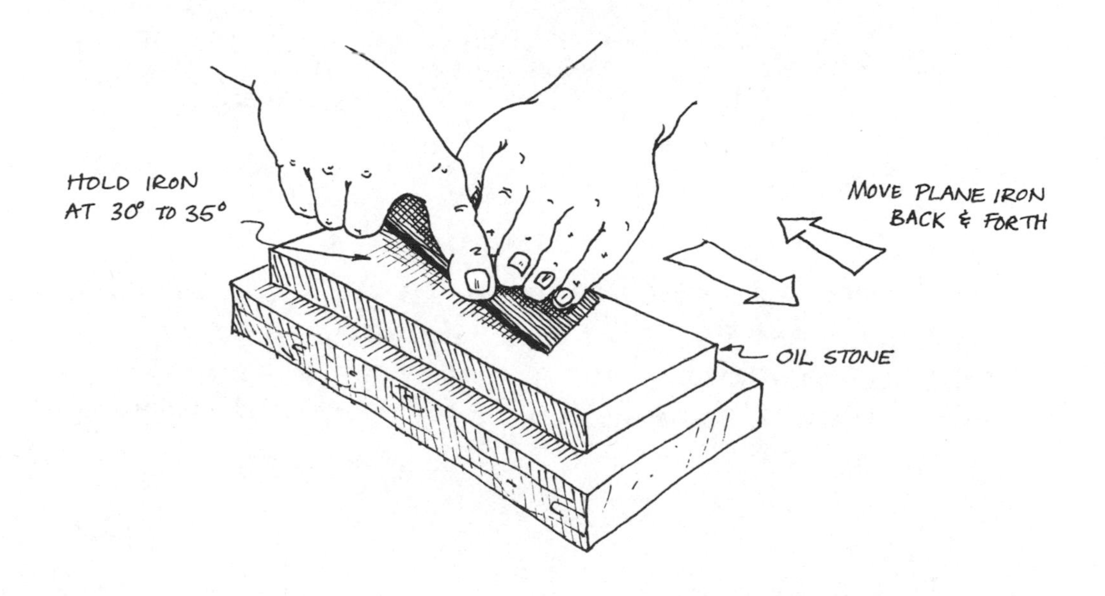

In order not to alter the angle of the bevel, you must take great care to hold your hands steady. After a while, a wire edge will develop on the back side of the plane iron. This must be removed by rubbing the plane iron on the **oilstone** whilst holding the iron perfectly flat on the **stone**. You must not make a bevel on the back side of the plane iron.

Try to wear the **oilstone** evenly, and keep the surface of the **oilstone** covered with a thin film of a lightweight oil. The purpose of the oil is to float away the particles of metal worn from the tools and keep the **oilstone** sharp.

3. CUTTING TOOLS **Chisels** are ground and sharpened just like plane irons, the only difference being that the angle of the bevel is different, and varies with the kind of **chisel**.

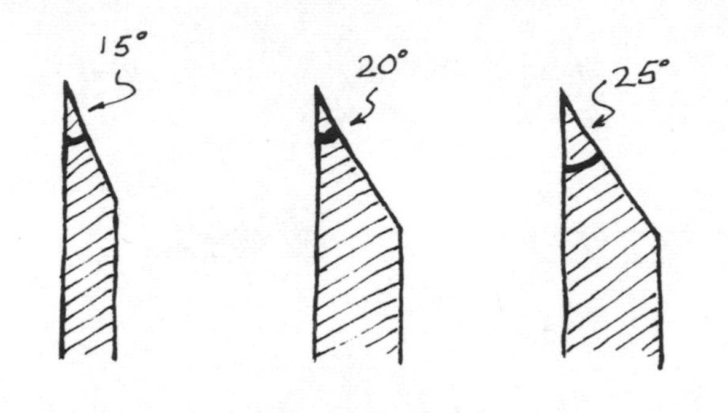

EXAMPLES OF THE VARIOUS BEVEL ANGLES ON DIFFERENT CHISELS

As with plane irons, remember to keep the cutting edge square, avoid beveling the flat side of the blade, keep the bevel flat, not convex, and if you grind, avoid overheating. If any of this should happen, however, you must start all over again from the beginning, and re-grind a new bevel.

In sharpening **knives** the same remarks on grinding apply, with the addition that you have more leeway in the matter of the shape of the blade. Examine different blades, they have different outlines for different jobs.

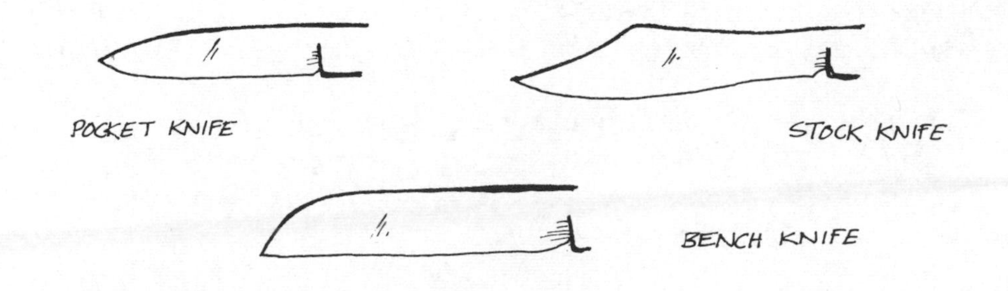

To whett a **knife**, place the blade flat on the **oilstone**, and then raise the back of the **knife** just enough to bring the cutting edge in contact with the **oilstone**.

1. KNIFE FLAT ON STONE

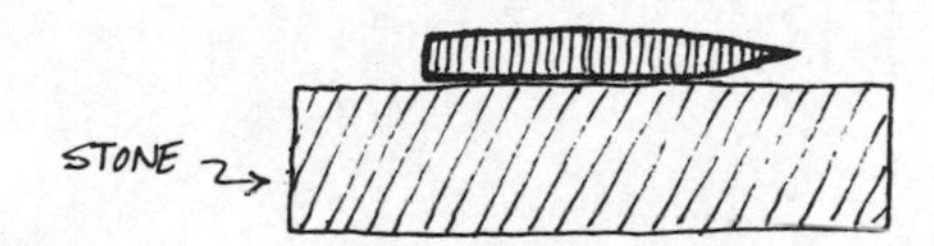

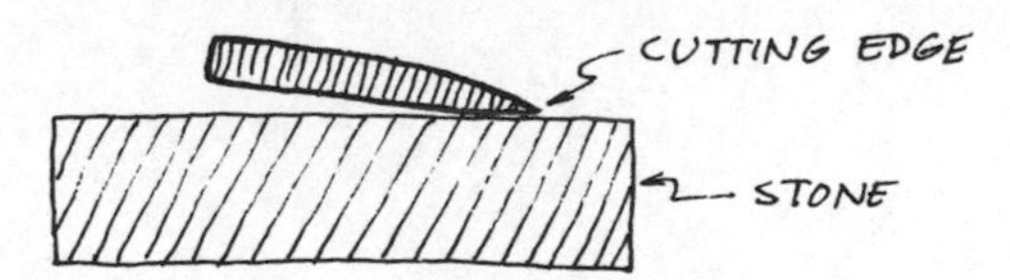

Then, applying moderate pressure, slide the blade across the surface of the **oilstone**, first on one side, and then on the other. Continue doing this, as illustrated below, until you have produced a wire edge that you can feel with your thumb. This you must remove by stropping the blade on a piece of leather.

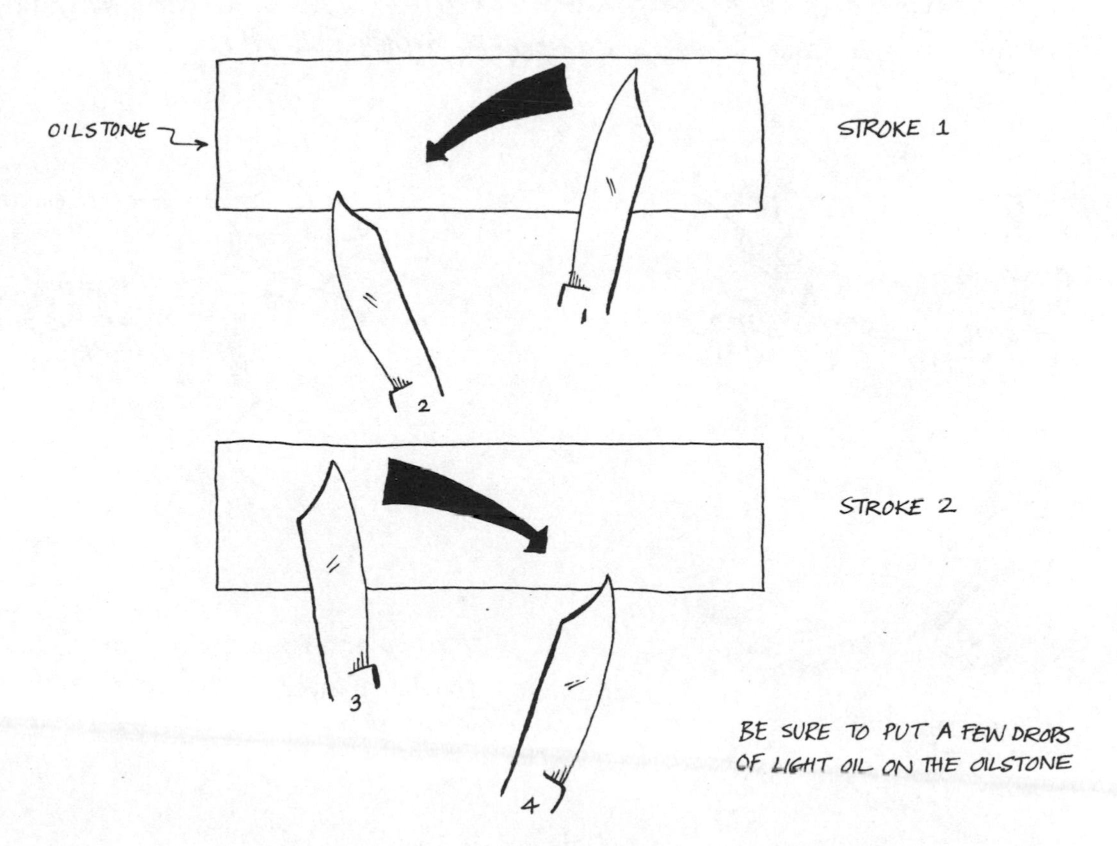

4. BORING TOOLS

Brad awls must be ground occasionally, a simple matter of grinding square across the tip.

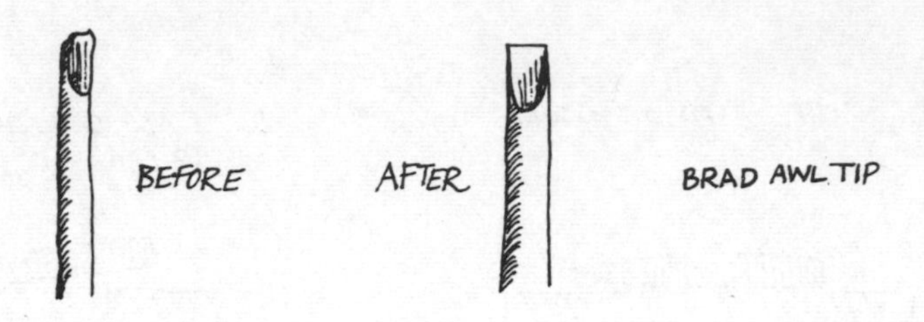

Hand drills and **braces** should be kept oiled and free from rust. Whereas **drill bits** are rather complicated to sharpen – as there are various critical angles to be maintained – **auger bits** can easily, and should be, sharpened. There are, in fact, only two points to remember. First, the spurs must be sharpened on the <u>inside</u> only; second, the cutting lips must be sharpened on the top side only.

Auger bits are best sharpened with a small three-sided **file**, although it is possible to obtain a special **auger-bit file**.

FILE THE SPURS HERE

FILE THE LIP, OR CUTTING EDGE, HERE

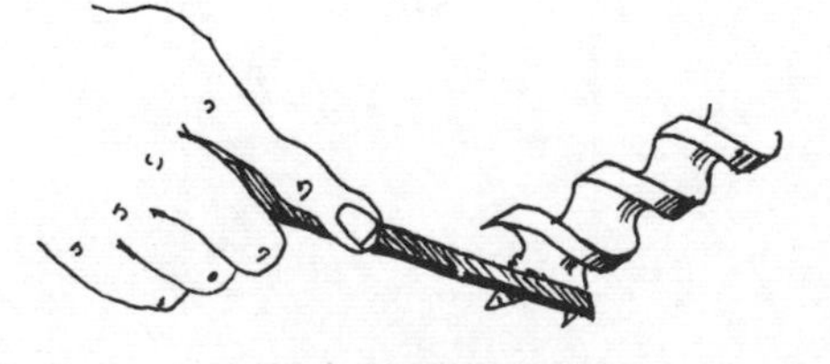

FILING THE TOP SIDE OF THE CUTTING EDGE

If a spur of an **auger bit** should become damaged, the **bit** will still work if you file it off completely.

5. STRIKING TOOLS **Hammers** and **mallets** do not really need sharpening, although sometimes it is possible to touch up the inside of the claw of a **hammer**, which may become damaged from pulling nails, as illustrated on page 88. However, do not be over-zealous in this regard, otherwise it will be diffiault to remove the pulled nail from the biting and sharpened jaws of the claws.

A chipped face on a **hammer** is dangerous, and such a tool should be discarded. Loose heads are similarly dangerous, whether it be a **hammer**, a **mallet**, or any kind of striking tool. The method of repairing loose heads is explained on page 95.

One last word: from time to time, rub all wooden parts of all tools lightly with linseed oil. It nourishes the wood and prevents it from drying out and splintering. Furthermore, as you handle the tools over a period of time, the linseed oil will help develop a beautiful patina.

Remember the two famous proverbs:

A BAD WORKMAN BLAMES HIS TOOLS!

YOU CAN ALWAYS TELL A CRAFTSMAN BY HIS TOOLS!

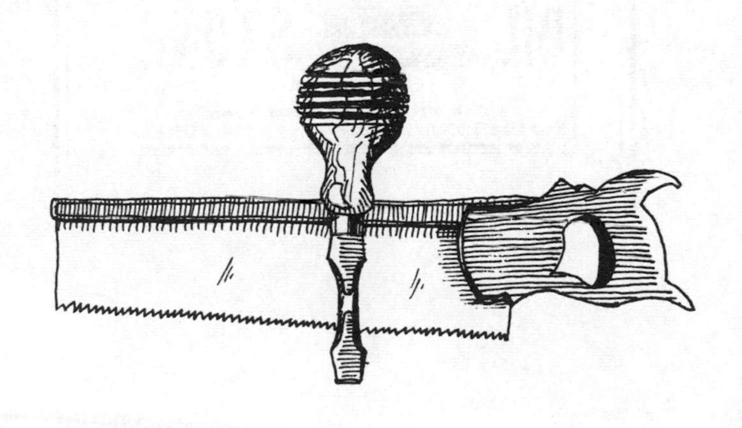

A. F. PIKE,

East Haverhill, New Hampshire,
Manufacturer and Wholesale Dealer in

Scythe, Axe, Knife and Hacker

STONES.

LETOILE,
UNION,
PREMIUM,
DIAMOND GRIT,
WHITE MOUNTAIN,
INDIAN POND (red ends)

Stones gotten up or labeled in any style desired. Price and quality guaranteed.

Our Stone are of good keen grit and *will not glaze.*

Steam Oil Stone Works.

F. E. DISHMAN,

Successor to WM. GALBRAITH & Co.
Manufacturer of and Dealer in the Best

Washita, Arkansas, Hindostan and Sand

STONES,

☞ Of various sizes and patterns, suited to every variety of Mechanical use. ☜ **New Albany, Ind.**
Send for price list.

BOYD & CHASE,

The largest manufacturers in the world of

OIL  STONE

Of all description.

107th Street and Harlem River,
Send for Price List. **NEW YORK.**

Bibliography

There are, of course, many good books on all aspects of woodworking, as the card index of any good library will show. However, I have felt that it would be useful to list some of the books which have helped me in the past. Not all of these are currently in print, but good books are sometimes reissued by concerned publishers, while others may be found with the help of book-finding establishments, second-hand bookstores, and perseverance, so I have thought it worthwhile to include them anyway.

HISTORICAL & BACKGROUND BOOKS :

Goodman, W.L. THE HISTORY OF WOODWORKING TOOLS. London : G. Bell and Sons, Ltd., 1964

Hibben, Thomas, THE CARPENTER'S TOOL CHEST. Philadelphia : J. B. Lippincott Company, 1933

Mercer, Henry C. ANCIENT CARPENTERS' TOOLS. Doylestown, Pennsylvania : The Bucks County Historical Society, 1951

Sloane, Eric, A MUSEUM OF EARLY AMERICAN TOOLS. New York: Wilfred Funk, Inc., 1964

Welsh, Peter C, WOODWORKING TOOLS, 1600-1900. Washington, D.C.: United States Government Printing Office, 1969

Wildung, Frank H, WOODWORKING TOOLS AT SHELBURNE MUSEUM. Shelburne, Vermont: The Shelburne Museum, 1957

BOOKS ON VARIOUS WOODWORKING TECHNIQUES :

Aller, Doris, SUNSET WOODCARVING BOOK. Menlo Park, California: Lane Magazine & Book Company, 1951

Boison, J, INDUSTRIE DU MEUBLE. Paris: Dunod, Éditeur, 1922

Griffith, Ira Samuel, ESSENTIALS OF WOODWORKING. Peoria, Illinois: Manual Arts Press, 1922

Haines, Ray E., Adams, John V., Van Tassel, Raymond, and Thompson, Robert L, CARPENTRY AND WOODWORKING. New York: D. Van Nostrand Company, Inc., 1948

Hayward, Charles H, CABINET MAKING FOR BEGINNERS. New York: Drake Publishers, Inc., 1974

Hodgson, Fred. T, MODERN CARPENTRY: A PRACTICAL MANUAL. Chicago: Frederick J. Drake & Co., 1909

Milton, Archie S., Wohlers, Otto K, A COURSE IN WOOD TURNING. Milwaukee, Wisconsin: The Bruce Publishing Company, 1919

Moxon, Joseph, MECHANICK EXERCISES. London, 1703 2nd edition by Praeger Publishers, Inc., New York, 1970

Nicholson, Peter, MECHANICAL EXERCISES. London: J. Taylor, 1812

Perth, L, THE STEEL SQUARE. New Britain, Connecticut: Stanley Tools, 1949

Popular Science ed., WOODWORKER'S TURNING & JOINING MANUAL. New York: Popular Science Publishing Co., Inc., 1934

Sutcliffe, G. Lister, ed., THE MODERN CARPENTER, JOINER, CABINET-MAKER. London: The Gresham Publishing Co., 1900

Ulrey, Harry F, CARPENTERS' AND BUILDERS' LIBRARY. Vols. I-IV. Indianapolis: Howard W. Sams & Co., 1970

ELEMENTARY CARPENTRY AND JOINERY. London: Ward, Lock And Co., 1870(?)

The following firm of specialist booksellers publishes a most comprehensive catalog of books on woodworking which they will send post free.

Stobart & Son, Ltd.,
67/73 Worship Street,
London, EC2A 2EL
England

The catalog is called: BOOKS ON WOODWORKING, METALWORKING AND HANDICRAFTS, and lists nearly 500 books.

Another catalog worth obtaining is that of

Woodcraft Supply Corp.,
313 Montvale Avenue,
Woburn, Massachusetts, 01801

One of the best woodworking tool supply companies, their catalog also lists many books on all kinds of woodworking.

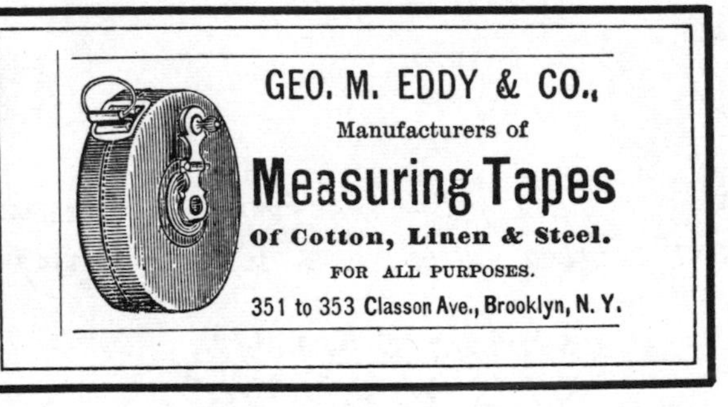
GEO. M. EDDY & CO.,
Manufacturers of
Measuring Tapes
Of Cotton, Linen & Steel.
FOR ALL PURPOSES.
351 to 353 Classon Ave., Brooklyn, N. Y.

Index

Adze, 59
Auger, 76
Auger bit, 76, 77, 81
    sharpening, 108
Awl, 17, 74
Axe, 59, 60

Back saw, 33
    use of, 41
Bar clamp, 29
Barrel hatchet, 60
Beetle, 86
Bell-face hammer, 87
Bench chisel, 64
Bench hook, 29
Bench knife, 106
Bench plane, 44-45
    use of, 53-56
Bench rule, 16
Bench stop, 28
Bench vise, 27
Bevel, 19
    use of, 27
Bit brace, 73, 76
Bit brace cont.
    use of, 81-83
Block plane, 45
    use of, 57
Brace, see Bit brace
Brad awl, 74
    grinding of, 108
Breast drill, 75
    use of, 80
Broad hatchet, 60
Buck saw, 32

Carborundum stone, 100
Carpenter's nut auger, 76
Carpenter's pencil, 17
Carpenter's rule, 16
C clamp, 29
Chain saw, 32
Chalk line, 18
    use of, 25
Chisel, 59, 62, 64-66
    bench, 64-65
    firmer, 64
    floor, 65

Chisel cont.
- framing, 64
- gonge, 66, 70
- mortess (mortis), 70
- paring, 64, 70
- sharpening, 106
- short-bend woodcarving, 66
- slick, 65
- socket, 65
- special-purpose, 65
- tang, 65
- turning, 66
- use of, 67-71
- woodcarving, 66

Clamp
- bar clamp, 29
- c clamp, 29
- wood screw, 29

Claw hammer, 85, 87
Claw hatchet, 60
Clinching, 92
Cloth tape measure, 14
Commander, 85, 86
Compass saw, 35
Coping saw, 35
Crosscut hand saw, 33
- use of, 36-38

Dado plane, 46
Depth gauge, 79
Dovetail joint, 34
Dovetail saw, 34
Drawknife, 59, 61
Drill, 75
- breast, 75
- hand, 75
- push, 75
- use of, 77-80

Drill bit, 75, 77

End grain, 57, 70
English hatchet, 60

File, 100, 101, 103
Firmer chisel, 64, 70
Floor chisel, 65
Floor plane, 51
Fore plane, 44
Former chissel (firmer chisel), 70
Four-fold two-foot rule, 16
Frame saw, 32, 35
Framing chisel, 64
Framing hammer, 85
Framing square, 20
Fret saw, 35

Gimblet, see Gimlet
Gimlet, 74-75
Gouge, 66, 70
Grain, 54
Grinding, 100, 104
Grindstone, 100
Grooving plane, 46

Halved-together joint, 9
Hammer, 85, 87
- bell-face, 87
- claw, 85, 87
- framing, 85
- jeweler's, 85
- joiner's, 93
- ladies', 93
- parts of, 87
- plain-face, 87
- sledge, 86
- tack, 85
- use of, 88-92

Hand drill, 75

Hand drill cont.
  parts of, 77
  use of, 78-80
Handle tightening, 95
Hand saw, 33
Hatchet, 59, 60
  barrel, 60
  broad, 60
  claw hatchet, 60
  English, 60
  lath, 60
  shingling, 60

Iron plane, 47

Jack plane, 44
Jeweler's hammer, 85
Jig saw, 32
Joiner's hammer, 93
Joint
  dovetail, 34
  halved together, 9
  mortise-and-tenon, 34
  open, double, mortise-and-tenon, 11
  open mortise-and-tenon, 10
Jointer plane, 45

Kerf, 37, 40, 41
Keyhole saw, 31, 35
Knife, 17, 59, 61
  bench, 106
  drawknife, 61
  pocket, 61, 106
  sharpening of, 106-107
  sloyd, 17, 61
  stock, 106
  use of, 62, 64
Knife cont.
  utility, 61

Ladies' hammer, 93
Lath hatchet, 60

Mallet, 69, 86, 87
  use of, 94
Marking awl, 17
Marking gauge, 18
  use of, 24
Maul, 86
Moon, 89
Mortess-chissel (mortise chisel) 70
Mortise, 34
Mortise-and-tenon joint, 10, 11, 34
Moulding plane, 46

Nailing, 88-93
Nail removal, 88
Nail set, 93

Oilstone, 100, 105
Old woman's tooth, 46
One-man crosscut saw, 32
Open, double, mortise-and-tenon joint, 11
Open mortise-and-tenon joint, 10

Paring chisel, 64, 70
Pencil, 17
  use of, 23
Plain-face hammer, 87
Plane, 43-57
  adjusting, 48-52
  bench, 44-45

Plane cont.
block, 45
dado, 46
floor, 51
fore, 44
grooving, 46
iron, 47
jack, 44
jointer, 45
moulding, 46
parts of, 47
plough, 46
rabbet, 46
sharpening, 104
smooth, 44
special-purpose, 46
wooden, 47
Plane iron, 48
Plough plane, 46
Pocket knife, 61
Push drill, 75
use of, 80

Rabbet plane, 46
Rip saw, 33
use of, 39-41
Rule, 15-16
graduations of, 20
marking with, 22
measuring with, 21-22

Saw, 31-35
back, 33
buck, 32
chain, 32
compass, 35
coping, 35
crosscut hand, 33
dovetail, 34

Saw cont.
filing, 103
frame, 32, 35
fret, 35
hand, 33
jig, 32
jointing, 101
keyhole, 31, 35
one-man crosscut, 32
rip, 33
set of teeth of, 39, 102
sharpening of, 101-103
table, 32
teeth, 39
tenon, 33
turning, 32
two-man crosscut, 32
Saw set, 102
Scratch awl, 17
Scriber, 17
Sharpening stones. see Oilstone
Shingling hatchet, 60
Short-bend woodcarving chisel, 66
Sledgehammer, 86
Slick, 65
Sloyd knife, 17, 61
Smooth plane, 44
Socket chisel, 65
Special-purpose chisel, 65
Special-purpose plane, 46
Sprig tool, 74
Steel tape measure, 14
Stock knife, 106

Table saw, 32
Tack hammer, 85
Tang chisel, 65
Tape measure, 13-15
Tenon, 34
Tenon saw, 33

Toenailing, 92
Try square, 19
  testing the accuracy of, 99
  use of, 25-26
Turning chisel, 66
Turning saw, 32
Two-man crosscut saw, 32

Undertaker's screwdriver, 109
Utility knife, 61

Vise, 27, 28

Wedge, 95
Whetting, 105, 107
Wimble, 74-75
Woodcarving chisel, 66
Wooden plane, 47
Wood screw, 29

Zig-zag folding rule, 15

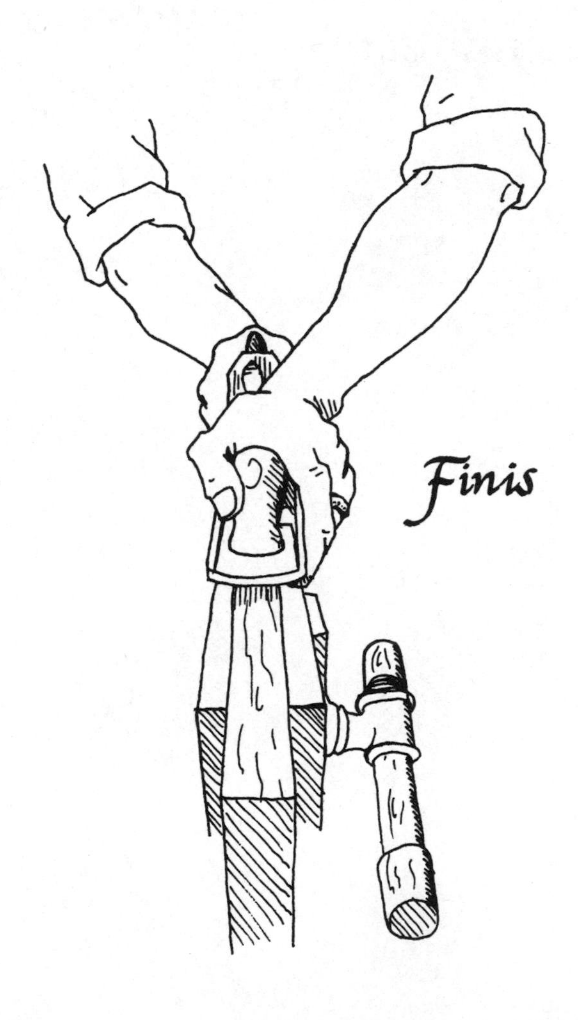
Finis